Smart Video Security Handbook

A Practical Guide for Catching Intruders Before They Act

By

John Romanowich and Danny Chin

with Thomas V. Lento

Published by Video Valley Press
745 Alexander Rd
Princeton, N J08540

ISBN-13: 978-0692536346 (Video Valley Press)
ISBN-10: 0692536345

First edition.

Table of Contents

Preface: A Practical Approach to Site Security

What you're reading is the culmination of years of efforts and countless hours questioning, debating, studying and developing solutions in order to solve one of the most elusive security challenges of all time: How can technology be used to dependably detect intruders and prevent harm to businesses, government facilities and critical infrastructure?

The Smart Video Security Handbook was written to meet the growing need for a concise guide to planning, selecting, and installing a smart video-based security system. Policy makers, site managers, security directors, architects and engineers, and installers can use it as a roadmap to create systems that deliver instant awareness of unauthorized intrusions and promote rapid response to threats.

Put simply, the problem of security has always been about human attention. It's just not possible for people to see all places at all times under all conditions. Smart cameras, properly deployed, focus people's attention precisely where and exactly when an intrusion occurs in a way that's repeatable and affordable.

As the founder and CEO of SightLogix, I can tell you firsthand that designing such a system was incredibly complicated, even though the solutions themselves appear to work so simply. It required engineering innovations and technology breakthroughs that seemed impossible only ten years ago.

> "Put simply, the problem of security has always been about human attention"

Most importantly, it's taken a team of dedicated, remarkable, tireless people who share my passion to make the world a safer place. Without them, and without the support of my family, investors, and the many friends I've made in the industry, this book and the solutions described within would never have been possible.

The outcome of all this effort gives every indication that it will be the next billion-dollar market. Smart video is definitely an idea whose time has come. Incidents of theft, sabotage, and even the occasional terror threat have afflicted public spaces, corporate installations, and vital infrastructure for years.

More recently there has been an alarming rise in the boldness of the perpetrators. Even when video cameras are plainly visible, they simply ignore them, certain that no one is watching and they won't get caught.

Older video surveillance systems actually do little to protect such assets. They are mostly limited to forensic recording and are passive in nature. Their primary use is in forensic analysis after an intrusion event.

Smart video security, by contrast, is an active, pre-emptive tool for thwarting intruders. It detects potential security breaches and gives security staff the visual and location data necessary to intervene, with very few nuisance alerts. In addition to perimeter protection, it will effectively cover specific objects and areas on a site, enabling a volumetric approach to security.

Nor is it limited to public and corporate applications. As smaller, less costly cameras appear in the market we can expect to see outdoor site security expand into the consumer space. Homeowners will use it to safeguard the borders of their property or monitor specific areas such as swimming pools, day and night. In addition, smart cameras promise to replace PIR motion sensors for indoor security, delivering higher detection accuracy and more reliable performance in all markets.

The Smart Video Security Handbook covers all major aspects of today's most advanced video security systems. The handbook

- Outlines the advantages and limits of video-based security
- Provides security pros with the information they need to select the proper equipment and use it effectively
- Covers ways to design systems based on effectiveness, ease of implementation, and cost
- Offers guidance in setting up security zones to assure adequate site coverage and effective response to events
- Includes sample plans of site security systems, including diagrams and explanations, to serve as templates for creating video-based security systems
- Discusses the potential for deploying a smart video system as a central component of a comprehensive security plan

SightLogix and the Smart Video Security Handbook

SightLogix was founded over a decade ago, in the aftermath of the 9/11 attacks on the World Trade Center. Since then our technical staff has focused its industry-leading expertise on developing the world's most accurate and reliable smart video security systems, with a goal of minimizing nuisance alerts. The Smart Video Security Handbook incorporates many of the lessons we have learned about smart video technology and its practical applications.

Today's smart video is an ideal solution to the new challenges in site protection that confront security professionals. It outperforms older technologies by a wide margin. It often costs less. Installation is less disruptive, and the technology is highly reliable. These advantages are covered in depth later in this book.

For those who are worried about having to learn a totally new technology, I have some good news. The learning curve for smart video security is not very steep. The Smart Video Security Handbook makes it even flatter.

Readers should note that the equipment specifications and sample plans in this book are based on the capabilities of SightLogix products. We cannot guarantee that other makers' equipment will perform exactly like a SightLogix product. If you adopt one of our plans but use a different brand of product in place of a SightLogix component, be sure to test the substituted product to see how it will perform in your application, and adjust the plan accordingly.

John Romanowich
CEO, SightLogix Inc.

Executive Summary

Thwarting Terror, Stopping Theft: The Future of Outdoor Smart Video Security

In June 2002 one of the authors of this book, John Romanowich, toured the World Trade Center site in lower Manhattan along with several other leading video technology experts.

Not quite two years had passed since a terrorist attack had taken down the Twin Towers, killing nearly 3,000 people. The area was both a crime scene and an active construction zone. Romanowich and his colleagues had been invited by the Port Authority of New York and New Jersey to determine whether video surveillance could provide security for the work being done there.

Their answer was an unequivocal "No." The video technology of the day just wasn't up to the job. The visible-light cameras typically used in surveillance systems were useless in the dark, while the video from costly thermal night-vision cameras was blurry, indistinct, and difficult to interpret. Even the logistics didn't make sense: installing and maintaining the necessary wired infrastructure was too difficult and expensive.

Today their answer would be totally different. New thermal video technology has swept away barriers of cost, complexity, and compromised performance. It is already delivering comprehensive 24-hour protection at sites that are a lot more difficult to secure than the 2002 World Trade Center.

In fact the smart cameras described in this book now secure the National September 11 Memorial at the new World Trade Center.

Smart thermal video technology gives security professionals a tool for creating the outdoor equivalent of a burglar alarm. It can

- Detect and display unusual movement on a site at any time of day or night, in almost any weather
- Assess the threat potential of each incident
- Alert security personnel when a threat is credible
- Direct cameras with telephoto lenses to zoom in for a closer look
- Supply the geo-registered location of the threat so guards can reach it faster and stop intruders before they act

We are truly at the dawn of a new era in outdoor site security.

This advanced technology arrives when theft is rampant and the whole world is under threat of terrorist attack. Well-organized criminals carry out raids on poorly protected storage areas. Misguided fanatics driven by nationalism, ethnic hostility, or religious extremism are willing and able to travel across countries and across borders to commit acts of sabotage and terror. They regard buildings, public spaces, airports, and infrastructure as convenient targets of opportunity. And while less dramatic than terrorism, theft and vandalism at outdoor facilities are more common and lead to millions of dollars of lost revenue.

Smart thermal video security systems can thwart these plans. The reason: video is the ultimate human intrusion detector. Instead of a single point of information about an intrusion from a fence-mounted sensor, video covers the area with thousands of information sources, updated 30 times per second. No other security technology, including radar, can match its coverage, its detection reliability, or its accuracy in dismissing false alarms.

Unfortunately, very few outdoor sites are protected by such thermal video systems. Now that cost and complexity have both fallen, security professionals should move to implement the newer technology wherever security is a true concern.

It is our hope that this handbook will accelerate adoption of the newer technology. It contains detailed information and advice on all the steps necessary to create and implement a truly effective video security system, from strategic planning to equipment selection, site design, and installation.

Those topics will be covered in later chapters. Here we are concerned with presenting the basic goals and advantages of smart thermal video security. We will also point to new directions in smart visible-light video that promise to transform indoor security operations the way smart thermal technology has reshaped outdoor protection.

Security: Not Just How, But Who

We'll start with a fundamental truth: security is only as effective as the people who provide it. However smart the alarms, alerts, and video systems may be, they can't make the final judgment to call the police or otherwise intervene when there's an intrusion on a site. Only the on-site security staff or the staff in a central station monitoring office can do that.

Given this obvious fact, the primary goal of any security support system should be to make the staff more efficient and more effective. Unfortunately conventional video surveillance systems, the kind that millions of sites have used for decades, don't do much to achieve the goal. In some ways they even work against it by creating a false sense of security.

Conventional video systems put surveillance cameras at strategic locations around a site, and display their imagery on one or more video screens in the security office. A video surveillance system may offer basic video analytics, allowing it to alert the guards when there is unusual movement in one of the covered areas, and it usually records the video for later review.

This helps security personnel the way a pair of binoculars helps a sentry in a tower. There's no question that the sentry is better off with binoculars. He doesn't have to patrol the site constantly to find out what's going on, and if he spots something out of the ordinary, he can focus on it. He can even record what he sees in a duty log.

But his field glasses are passive – and so is surveillance video. Both extend the user's view, but neither makes an active contribution to the security function. That is, they do not help staff detect and analyze potential threats more efficiently and effectively.

Smart video security systems, on the other hand, are active by design. They deploy cameras with built-in intelligence. These cameras optimize the video, detect unusual activity in the video stream, and bring it to the attention of humans to analyze its threat potential. And they do this with high accuracy over very large areas. They are the equivalent of having extra agents in the field, screening information and only alerting the main security staff when what they find is suspicious and actionable.

Essentially, then, a smart video security system is a force multiplier, taking the burden of monotonous surveillance off regular security staff. Instead of just watching endless video feeds, the staff gets information that lets them do their jobs better.

That in a nutshell is the crucial difference between video *surveillance* and video *security*. Video *surveillance* is passive. Its presence does not significantly impact the way a security staff works. Video *security*, on the other hand, is active. It helps the staff work smarter, which enhances the security function at every level.

Dependable Alerts, Assured Response

Effectiveness and efficiency involve more than just working smarter. We also want to help our security forces respond quickly and dependably when there is a serious perimeter breach. As recent widely publicized break-ins at high-profile sites prove, quick response is not a given, even when an elite force guards a facility.

In 2014, for example, one man got his 15 minutes of fame when his trespass into the White House made every network newscast and cable news channel. He had managed to bypass Secret Service patrols, cross the White House lawn, gain entrance to the building, and penetrate as far as the Green Room. His progress was relayed to Secret Service guards in live video feeds, but nobody moved to stop him.

In 2012 a group of three protesters, including an 82-year-old nun, penetrated the heavily protected Y-12 National Security Complex. They spent two undisturbed hours inside the nation's largest nuclear materials storage facility before members of its 500-strong security force showed up.

There have been similarly troubling incidents at many government and military facilities over the past few years, and at airports, infrastructure installations, and other important and vulnerable locations. An Associated Press investigation found that intruders successfully evaded security at America's busiest airports at least 268 times between 2004 and 2014, getting on site, driving cars onto runways, even hopping onto jets.

Alert Fatigue

Who's at fault? In most cases the shortsighted response is to single out the security force as scapegoats. The guards were not alert. They did not respond to warning signs. They didn't take their job seriously. They and their bosses should be punished.

This rush to judgment ignores the real problem: alert fatigue. Security officers are charged with maintaining 24-hour vigilance about what is happening at their outdoor facility. After responding to hundreds of alerts about perimeter breaches that turn out to be nothing more than small animals or windblown branches, even the most conscientious security guards lose confidence in the system and start to ignore its warnings. That is simply human nature.

"Hundreds of alerts" is not an exaggeration. A recent study compared a conventional visible-light video surveillance system with analytics to a smart thermal video security system. The surveillance system posted 750 nuisance alerts during the study. At the same site and in the same time period, by contrast, the smart video security system registered only a handful of alerts, just enough to prove that it was on the job.

Clearly what we have called alert fatigue is a serious problem. You cannot protect an outdoor site if the security staff begins to ignore the intrusion warnings it receives. Systems that inundate guards with false alarms do not enhance their effectiveness – they diminish it.

> "After hundreds of alerts turn out to be small animals, it's simply human nature to start ignoring them."

More seriously, eventually one of the warnings will not be a nuisance alert. It will be a real breach. Will the guards respond?

They will if constant nuisance alerts have not undermined their confidence in the system's reliability. As this handbook will make clear, smart thermal video security systems promise to detect all intrusions, but ignore almost everything that isn't a threat. They don't flood the security staff with false alarms. They provide only actionable information, and help the staff verify the nature and scope of every breach.

Smart thermal video security is the obvious choice for outdoor site security applications in our modern world.

Practical Matters

Our objective in this handbook is to describe today's best practices in site security, and to provide a practical guide for designing and using video-based outdoor systems. Topics will include

- Advantages and limits of smart thermal video security
- How to select the proper equipment and use it effectively
- Designing systems for effectiveness, easy implementation, and low cost
- Ensuring maximum site coverage and prompt response to events
- Sample site plans for security systems, with diagrams and explanations
- Using smart thermal video with a security operations center (SOC) or central station (CS) monitoring service

Part I of the book introduces smart thermal video technology and explores its potential to transform site security. Chapter 1 will be of particular interest to policy makers and facility managers for its summary of where we are today in video security technology, and its description of the tools available to protect outdoor assets. It is essentially a case statement for upgrading facilities to a newer, far more effective technology.

Chapter 2 explores the technology of smart thermal video and its benefits when used to protect an outdoor site, while Chapter 3 covers questions of cost from system and operational perspectives.

Part II covers the nuts-and-bolts of planning and designing an effective video security system to detect intruders and alert security staff to threats. Chapter 4 covers the selection of cameras and other equipment and shows how to implement their various functions. In Chapter 5 readers will find model site designs to give them a starting point in building their own systems.

Finally, Chapter 6 is our attempt to look ahead at what the future holds for the people responsible for maintaining the security of outdoor sites and their assets. It covers corporate and public security, and foresees the application of smart video technology to the monitoring and protection of indoor spaces and residential and personal property as well.

There can be no doubt that the threats we face will change and grow over time. Video technology must evolve to counter new methods of attack. The good news is that the technology is as flexible as it is powerful. There's a lot of room left to develop sophisticated new functions as countermeasures against those who would do us harm.

Part One

Effective Site Security: The Smart Video Solution

1

Outdoor Security Comes of Age

Theft, sabotage, and terror attacks are on the rise at outdoor sites around the world. They target transportation infrastructure, utility installations, corporate and governmental buildings, public gathering places, and more.

Whatever the motivation for these attacks, from fanaticism to greed, they pose a clear threat to economic progress, social stability, and public safety.

To cite just one example on the economic front, copper thefts from utilities in the US have more than tripled over the past five years. Losses are running at up to $1 billion per year. In one case a group of thieves made off with $68,000 worth of copper wire in 1,000-pound rolls from an outdoor storage space. A stunt like that takes determination and considerable resources.

Politically-inspired actions are even more worrisome. This book's Executive Summary referred to a 2012 break-in at the Y-12 National Security Complex at Oak Ridge, TN. Three peace activists, one of them an 82-year-old Catholic nun, cut through three fences, avoided dogs and guards from a 500-person security force, and made their way to a storage bunker containing much of America's bomb-grade enriched uranium.

They banged on the bunker walls with a hammer, splashed them with blood, spray-painted slogans, hung antiwar banners, and sang protest songs until the guards finally arrived – two hours after the break-in. Had the intruders been intent on an act of terror or mass destruction the consequences could have been catastrophic.

> "Conventional measures produce hundreds of alarms per month. That's not protection; it's chaos."

Why didn't Y-12's elaborate security precautions protect it? One fact stands out. Y-12's security officers admitted afterwards that their electronic intrusion sensors and video surveillance systems produced so many false alarms that they had stopped paying attention to them.

If that sounds irresponsible, consider the circumstances. Security systems using conventional measures, including video surveillance cameras with motion detection, produce hundreds of alarms per month. This is not protection; it's chaos. No wonder the guards stopped responding.

There is no longer any reason for this situation to exist. This chapter relates how smart thermal video security technology, the successor to video surveillance, can effectively protect outdoor assets and help security forces stop intruders before they act. Systems using this technology

- automatically and accurately detect, assess, and locate threats to outdoor assets
- display a thermal video image of the suspicious activity
- alert security personnel to the potential problem
- tell them where it is
- direct a visible-light camera to zoom in on the problem for a better view

Even better for those with limited resources, the superior performance of smart thermal video security solutions comes at an affordable cost.

In the following pages we will look at the key features of smart thermal video security and discuss its advantages in real-world outdoor applications. We believe the case for making smart thermal video security the centerpiece of any strategy for protecting outdoor sites will prove overwhelming.

Later chapters will address the technology behind the system, and provide tutorials on the best ways to implement it.

A New Model for Site Protection

It's relatively easy to protect an indoor space. Set your burglar alarm, lock the doors behind you, and go about your business. Door, window, and motion sensors will detect any break-in and either summon a guard to take action or notify a central station (CS) monitoring service to call for the police to investigate.

Outdoor sites pose a much bigger challenge. Coverage areas are large and irregular, making deployment of sensors a challenge. Wind, rain, snow, and other environmental factors can interfere with detection. The brute-force solution of blanketing the area with guards doesn't make economic senses unless you're protecting a military base.

Conventional Video Surveillance

Video coverage promises to solve those problems. However, visible-light video and perimeter sensor systems, which still account for the majority of installations, do not create a new security model. They are simply an add-on to the existing one.

Such conventional video surveillance systems deploy one or more cameras around a site. They are designed to let security guards in a central location enhance the scope of their vigilance by observing remote areas of a site on (usually multiple) monitors, as shown in Figure 1.

Surveillance systems are essentially passive. Installing them is like giving binoculars to sentries in a tower: it extends their field of view, but does not fundamentally change how they detect or react to threats. Guards are still expected to keep unblinking watch over what they survey, either by eye or video feed.

Figure 1: Video Surveillance Monitoring

This is setting a security force up for certain failure. After just a few minutes of staring at screens even the most conscientious guard begins to suffer from fatigue. We now know that video monitors go unwatched most of the time.

Video surveillance systems are primarily useful as forensic tools. They record the on-screen activity, which may allow you to identify an intruder after the fact. But they don't play an active role in keeping an intruder from wreaking havoc at your site.

Video Security

More advanced video systems do create a new model for site security. They use video analytics to detect movement in their field of vision, then alert guards to the possible presence of intruders (Figure 2).

Figure 2: Motion detection in action

These are video security systems, as distinguished from the older surveillance systems. While surveillance setups are passive, the real-time detect-and-alert functionality built into the new security systems make them active contributors to the security function. They retain the forensic functionality of surveillance systems, but add the capability to initiate pre-emptive intervention in the event of an intrusion.

The drawback of visible-light video security systems, as we have noted, is their tendency to generate a flood of nuisance alerts. They are smart enough to detect moving objects in the video stream, but not smart enough to reliably distinguish between a human bent on mischief and an animal trotting through a site. With some visible-light systems an enterprising squirrel can set off a false alarm.

To complicate the issue, animals aren't the only things that move in an outdoor setting. In fact, everything moves, and everything can trigger an alert. Tree branches wave in the wind, clouds cast moving shadows, debris flies across the ground. The subtle swaying of a pole-mounted camera can be interpreted as movement too.

Another variable is whether the camera can see at all. Visible-light cameras are highly susceptible to the vagaries of weather and lighting. Fog, rain, and noontime glare can blind them. Reflections from puddles or even headlights passing through the scene can trigger alarms. At night, when most intrusions take place, the area must be evenly lighted to produce a good image and reliable detection, which can be costly.

Fortunately, the smart thermal video technology now available has none of these shortcomings. It fully realizes the promise of video security.

Smart Thermal Video Security

The image in Figure 2 above is obviously not from a conventional visible-light video camera. It was captured by a thermal imager, also called a long-wave infrared (LWIR) detector. Unlike visible-light cameras, which capture direct and reflected light from a scene to form a picture, thermal cameras generate images based on the differing levels of heat emitted from various objects in the scene.

Among their advantages:

- They "see" in the dark
- They work 24 hours/day with no need for costly artificial lighting
- They ignore reflections, shadows, moving headlights, direct sunlight, and other light-based phenomena that can trigger alarms in a visible-light system
- Because humans give off heat, thermal sensors are perfect "people detectors," far more effective in spotting a person than visible-light models
- They detect body heat as far away as 600 meters – a third of a mile
- A single thermal camera can protect an area the size of a football field
- Proper design makes them immune to the effects of weather and

other environmental factors

At this point it is important to note that the primary function of a thermal camera is detection, not creating a high-resolution, crystal-clear picture. In fact, early models presented smeary, low-resolution images. The image in Figure 2, captured at night, actually represents a marked improvement over those obsolete units. While it's not clear enough to resolve the many details of the figure and its surroundings, it is perfect for intruder detection purposes.

Since the primary purpose of a thermal camera is to detect motion around the clock, improved resolution and visual quality may seem like a negligible advantage. When the improvement is this dramatic, however, there will be a positive effect on both the range of the camera and its motion detection performance, as discussed in Chapter 2.

Smart Thermal Video: Securing the Area

To reiterate, by replacing visible-light cameras with the latest smart thermal models you turn a marginal asset into a powerful aid in protecting an outdoor site from intruders bent on mischief. Their heat-based imaging makes them much better at detecting suspicious activity, and their potential for sophisticated analysis and reporting is something you won't find with visible-light units.

This doesn't mean that visible-light cameras are of no use. In fact, they are an important adjunct to smart thermal cameras within an integrated security system. When a thermal camera detects suspicious movement, it can direct a pan-tilt-zoom (PTZ) visible-light camera to take a closer look at the activity and display it in full-color HD video for positive identification.

Later we will look at how performance differences among various thermal camera models affect system effectiveness. Our concern here is with the bigger picture. What is the unique value proposition of a smart thermal video security system? In what applications has the technology been proven? How does it fit within an overall site security strategy?

Force Multiplier

A well-designed thermal video security solution fulfills the primary goal of any security system: improve the efficiency and effectiveness of the security force. It feeds guards reliable, accurate information, alerting them when there is a credible threat of unauthorized intrusion. Instead of being occasional watchers of continuous video feeds, guards become analysts and fast-response specialists who deal only with actionable situations.

Smart thermal systems also act as force multipliers. For example, rather than having one guard sitting in a monitoring station and another on patrol, you can have the cameras feed information wirelessly through the cloud to an iPad or smart phone in a patrol car. The roving guard doubles as a mobile monitoring station. That is impractical with conventional security systems, due to their high level of nuisance alerts and the complexity of the task.

Only a human observer can make the ultimate determination on whether unusual movement or suspicious behavior within an outdoor site constitutes a threat. If an apparent breach turns out to be real, only a security guard can neutralize it. By installing a smart thermal video security system you enable the security force to do that job better.

Strategic Implications

Thermal video offers a quantum leap in accuracy and reliability over previous approaches. Such systems already protect thousands of different outdoor sites around the world.

Security professionals working to protect assets from theft or the threat of vandalism, sabotage, or terrorism have found smart video security to be an effective answer to the threat of intrusion and disruption across a wide spectrum of vital economic and social areas. A few examples will indicate how well it adapts to different situations.

Energy Infrastructure

Transformer substations and other unguarded components of the electrical grid are especially vulnerable to trespass. The billion-dollar annual loss from copper theft in the US is one instance. But utilities also face an increasing threat of vandalism and sabotage.

This threat is being taken seriously. After an April 2013 incident when gunmen with .30 caliber sniper rifles caused over $15 million in damage to 17 transformers at Pacific Gas and Electric's Metcalf Transmission Substation outside of San Jose, California, the Federal Energy Regulatory Commission (FERC) ordered the North American Energy Reliability Corporation (NERC) to draft the CIP 014 Critical Infrastructure Protection Standard.

NERC CIP 014 requires transmission owners and operators to identify and protect critical transmission stations, substations, and control centers. A growing number of these facilities in the US and abroad are installing smart thermal video security systems to provide intrusion detection with a minimum of nuisance alerts.

Petroleum refineries, particularly those in countries experiencing unrest, are also using smart video as a safeguard against intrusion by people bent on mischief, damage, and destruction. As a bonus it can provide early warning of overheating or fire.

Transportation

Smart thermal video security is especially effective in addressing the complex security situations found in airports. In tests by the Transportation Security Agency (TSA) at the Buffalo Airport, a smart thermal system

- Successfully detected all 900 staged intrusions
- Sounded an alarm each time
- Displayed video of the problem
- Provided information on location, nearest camera reference numbers, and date and time
- Controlled PTZ cameras to zoom in on and follow detected targets

A public report on the tests is available at http://bit.ly/tsa-sightlogix.

Mass transit and railway systems also benefit from smart thermal video security in the station and beyond. Systems have been developed to detect people on or around train and subway tracks in some of the world's most critical rail facilities.

In March 2012 the New York State Bridge Authority (NYSBA) installed sophisticated smart thermal video security systems on five of its bridges across the Hudson River. When a vehicle or boat that stops for too long or moves too slowly, the systems not only alert security staff, they provide information on the object's size, its exact location (through GPS registration), its direction, and its speed. The expansive, unfenced areas alongside the bridges are also secured against pedestrians approaching the structures. If this data indicates a security problem – someone preparing to trespass for whatever reason, including sabotage – officers can be dispatched to handle it.

Government and Commercial Structures

Smart thermal systems provide 24-hour awareness of the approaches to courthouses, banks, data centers, and other vital buildings that terrorists may target.

Effective Protection for Outdoor Assets

Smart thermal video technology is creating zones of security where none existed before, operating effectively in highly complex environments such as airports and train yards. It is reliable enough for central station (CS) monitoring, and flexible enough to be integrated with security operations center (SOC) functions.

In short, a smart thermal video system should be part of any future plans for outdoor site security.

What about Indoors?

Smart thermal video security has proven itself to be the most advanced technology available for detecting human intrusions onto an outdoor site with low nuisance alerts. Indoor security systems may be on the threshold of a similar leap forward.

Indoor systems typically depend on door and window alarms, and on motion sensors inside the facility, to detect intrusions. But recently thieves have managed to get past these systems to steal tens of millions of dollars' worth of drugs and other valuable goods from industrial warehouses, offices, and other buildings.

A volumetric sensor – a smart video camera – would make these break-ins much more difficult, if not impossible. The technology can be adopted from the outdoor systems for which it was created.

At this point the chief impediment to the use of a smart camera for indoor security is cost. Cameras of any kind are more expensive than the motion sensors they would replace. Smart video cameras are even more costly, with thermal models the highest priced of all.

That may soon be less of an issue. For one thing, with the addition of cost-effective LED lighting, the less expensive visible-light cameras are adequate for security purposes at many indoor sites. For another, the relentless downward movement in the cost of virtually all electronics is already reducing the price of the cameras and of the smart processing technology that makes their intrusion detection so reliable. We're at the point where it no longer makes sense not to have a smart video security system covering vulnerable indoor sites.

We will revisit the question of indoor security in Chapter 6, which speculates on the future of smart video-based security. Before that, however, we must take a closer look at how smart video technology achieves its high level of performance.

2

Fail-Safe: The Smart Thermal Video Advantage

Visible-light video cameras, combined with perimeter intrusion detection sensors, have been standard equipment in outdoor security applications for decades. When used to detect intruders, these systems are not only expensive, they are so prone to nuisance alerts that security officers sometimes shut them off. It's only in the past few years that thermal video technology has matured to the point that it has begun to displace this old model of site protection in a big way.

This chapter will explain how smart thermal cameras work, and point out which of their features or functions should be on the security system planner or manager's checklist. Our discussion will cover

- How the technology solves specific outdoor security issues
- Capabilities that enable a smarter, more reliable security system
- Customization options, such as policy-driven response to potential threats
- What to look for in a camera: performance, features, functions
- Integrating the cameras with other equipment, including visible-light IP Dome cameras

We will focus special attention on video signal processing, the "smarts" of a smart camera. Without it a thermal camera would be just a night-vision version of a conventional surveillance camera.

Smart Thermal Cameras: Clear Images and Automated Detection 24 Hours/Day

Smart thermal video security cameras offer a virtually fail-safe solution to the challenges of guarding an outdoor site. They can pick a person or vehicle out of the background under virtually any weather or lighting conditions, determine the target's location, and apply pre-determined policies for assessing the threat level before alerting security staff.

While early thermal security cameras were only useful at night, today's thermal cameras function equally well in the daytime. In fact they handle certain daylight scenes better than visible-light cameras, producing clear video in extremely bright sun or through fog, mist, and rain.

Thermal security cameras can outperform visible-light cameras because they do not rely on light to create their images. They work by detecting thermal or heat energy emitted by objects in a scene.

All objects emit heat to varying degrees. Some generate it themselves (warm-blooded mammals, automobiles), while others absorb heat from the sun or other sources and then release it when their surroundings cool (rocks, pavement, soil).

Humans, the real threat to site security, maintain a normal body temperature of 98.6°F. The imager in a thermal camera can detect a human's heat profile even in total darkness. It can then create video that represents the various heat profiles in a scene, including those of people, in monochrome grayscale shades ranging from white to black, much like a black-and-white movie.

Thermal cameras are especially good at "seeing" human intruders at night, when other objects on an outdoor site are usually cooler than people. A thermal image shows warmer objects as brighter than their surroundings, so people will usually stand out as bright objects. Conversely, on a hot sunny day, when surrounding objects such as buildings or pavement reach very high surface temperatures, humans will be presented as darker than their surroundings. Bright or dark, however, intruders will always be visible as separate objects in a smart thermal camera image.

Visible-light cameras, on the other hand, work like our eyes: they can only see objects that produce light or are illuminated by the sun or some other light source. As a result, nighttime surveillance by visible-light cameras is impossible without covering the area with costly artificial lighting.

The reason early thermal cameras were restricted to night-vision duty was that their images of daytime scenes appeared washed-out and blurry. They also exhibited diminished performance at night when environmental conditions such as rain, heat, and humidity affected their ability to distinguish objects.

Figure 3: Rain and humidity mask crucial image detail in this video frame from an older camera

In Figure 3, for example, rain has cooled the scene until there is very little temperature differentiation among the objects in the scene. It's obvious that the older camera that created this image had trouble distinguishing between the person and the background. The result: a blurry image that masks detail to the point that it nearly conceals the target, creating a potential gap in security.

Such poor performance is a relic of the past, thanks to major advances in thermal sensors (imagers) and image processing technology, as Figure 4 (below) demonstrates. Today's cameras also make it possible to put more pixels on a target for better detection.

Technology Realignment #1: Imagers

Current thermal (infrared) imagers deliver substantially better sensitivity than those of just a few years ago, yet cost much less.

High-sensitivity imagers can "see" objects with a lower heat signature, such as those that are farther away. They are also able to sense temperature variations of 1/20th of a degree or less. This allows them to discriminate among objects when rain, humidity, or other conditions have reduced the temperature differences across a scene.

Smart thermal video cameras that take full advantage of these technical advances can detect potential threats in a scene around the clock, even when working under adverse conditions such as rain, snow, or mist.

Camera Selection Guide: Imagers

To get the highest quality raw imagery for your video security system, compare the performance of several cameras.

- Look for models that produce the highest-contrast images with the clearest differentiation among individual objects, under all conditions
- Give as much weight or more to signal processing (described below) as to imagers and lenses.

The last point is especially important. Thermal cameras need powerful signal processing capabilities if they are to provide reliable intruder detection.

Technology Realignment #2: On-Camera Image Processing and Detection

If its imager is a thermal camera's eyes, image processing is its brain. It is image processing that turns what is essentially a passive surveillance tool into an active security device, detecting motion and recognizing targets under any conditions.

Outdoor video security systems face a constantly-changing environment that often interferes with their ability to detect potential threats. Cameras mounted on poles shake and sway in the wind. Clouds create moving shadows on the ground. Foliage waves in the breeze, adding more movement to the scene. Rain, snow, and blowing dust occlude a camera's field of view.

Distinguishing legitimate targets amid all that noise is a significant task. To give some idea of the scope of the challenge, consider just the first step: making the scene visible to humans.

Thermal cameras can capture 16,384 shades of gray, from which they create a finely gradated image that clearly distinguishes one object from another. But humans can see at best only 256 shades of gray.

Converting the rich thermal imagery to 256 shades for human vision requires considerable processing power. A 640 x 480 video image contains 307,200 pixels. Each pixel has 14 bits of data, which equates to the 16,384 shades of gray mentioned above. These images are updated at a rate of thirty frames per second to present a smooth video image. The math is daunting:

640 x 480 x 14 x 30 = 129,024,000

In sum, cameras must process over 129 million bits of data every second, day and night, to present imagery that human eyes can resolve. And that's just the first step in detecting threats.

Intelligence at the Perimeter

It's natural to assume that this volume of processing is best done on a husky computer located in the central security office. But in fact you're much better advised to assign the task to powerful digital image processors built right into the cameras.

Using local processors lets you avoid having to compress the data before it reaches the computer. Sending 129 Mb/s of camera data through a network for processing virtually requires that the data be compressed. That's a problem because compression removes most of the finer scene details – up to 99% of the original data – seriously degrading a security system's ability to accurately detect and recognize targets in the video image. In fact, on days with restricted visibility due to rain or fog, data compression has caused processors to miss virtually all suspicious movement in a scene.

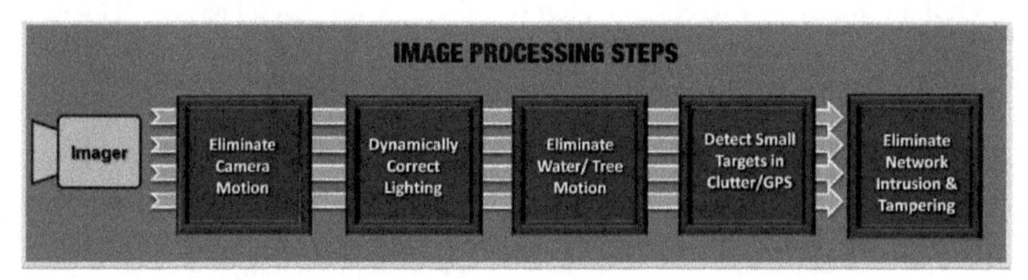

Figure 4: Smart camera with digital image processing

When the uncompressed imagery is processed in the camera, 100% of the raw scene data is available for analysis. With on-board image processors examining the full visual detail of every video frame, you can achieve a much greater degree of accuracy in detecting motion and recognizing potential threats.

Figure 4 illustrates the on-board image processing functions. The first processing step cancels out any camera movement embedded in the video. This is known as image stabilization, a more robust version of the "anti-shake" feature found in consumer still and video cameras.

When a camera sways on a pole, or shakes from vibration transmitted through its mounting hardware, its vertical, horizontal, and rotational motion appears in the video. Electronically stabilizing the image at the outset prevents the video analytics engine from interpreting this motion as suspicious movement in the scene.

We will look at the other important functions performed by the image processing in Figure 4 little later. Meanwhile, here are some tips on selecting cameras based on data processing and image stabilization.

Camera Selection Guide: Processing

Take these factors into account when choosing cameras:

- Limit your choice to models with high performance built-in processing to avoid data loss. There are big differences among manufacturers in the processing power and sophistication of their image analysis
- Look for cameras with excellent image stabilization performance, to keep camera movement from affecting video analytics
- Narrow the selection down to cameras that produce clear video under the most demanding environmental conditions (fog, rain, total darkness)

Reliable intruder detection requires processing 100% of video data. High performance on-camera processors are the surest path to that goal.

Technology Realignment #3: More Pixels on the Target

Smart video security systems enable three tasks: reliable detection of movement; recognition of potential targets as human, animal, vehicle, etc. plus automated tracking of targets; and possibly the identification of a specific object (who the person is, license plate numbers on a vehicle).

Each task requires that the target occupy a sufficient number of pixels in the camera image – that is, it must fill a large enough segment of the scene – to ensure accurate analysis. It takes more pixels to recognize a target than it does to detect movement, and even more to positively identify the target.

In practice, for example, reliable motion detection occurs when the moving object occupies about 24 or more total pixels of the image.

Recognition-level video needs at a minimum:
- 20 horizontal pixels on target (60 pix/m)
- 115 vertical pixels on target

Identification-level video needs at a minimum:
- 75 horizontal pixels on target (150 pix/m)
- 250 vertical pixels on target

How many pixels an object fills depends on five factors: the object's size, its distance from the camera, the focal length (magnification) of the camera's lens, the physical size of the imager, and the camera's resolution. A properly designed multi-camera security system will be laid out so that at least one of the cameras will put enough pixels on any object of interest anywhere in the site to detect motion and recognize the target.

Camera Selection Guide: Pixels on Target

Build your smart thermal video security system around the cameras that give you the most cost-effective coverage of your outdoor site.
- Larger sites should be equipped with models with a long detection range. You will need fewer of them, making them a cost-effective choice.
- If the site is small, consider lower-resolution cameras that still put enough pixels on a target for reliable recognition given the size of the area. Cost-per-camera will be lower, and since you won't need many for full coverage of the smaller area, you save money.
- When monitoring an area, choose wide-coverage thermal cameras with higher resolutions. One such camera can cover an area that would require up to 10 visible-light units.
- Smart thermal cameras are generally used for detection and recognition. Use a high-resolution PTZ (pan/tilt/zoom) or IP Dome visible-light camera for identification. A properly equipped smart thermal camera can send GPS coordinates to the PTZ to point it right at a recognized threat.

Operation, Integration, Implementation

Image stabilization, described above, removes camera motion artifacts from the video signal. It gives the other image processors in a smart thermal camera (as in Figure 4) a solid, steady image to work with.

Image processors downstream from the image stabilizer carry out two functions: image enhancement and video analytics. Video analytics processors have the glamor job – they detect motion, analyze the moving object, and send out alerts when necessary. The image enhancers prepare the video for human visualization.

Image Enhancement

Figure 5 graphically demonstrates how intelligent image processing by in-camera processors can bring clarity and realism to a subpar video of a daylight scene. The image on the left comes from a thermal camera without image enhancement. The image on the right was captured by a thermal camera with built-in image processing.

Both images were captured in bright daylight at the exact same time after objects in the scene had absorbed considerable heat from the sun. Due to a phenomenon known as thermal loading, there are so many hot objects and surfaces that the camera without processing capabilities cannot discriminate among them. Its image displays extensive white-out, and much of the detail in the scene has been lost.

Figure 5: Comparison of thermal images without and with image processing

Looking at the image on the left it's easy to understand why thermal cameras were once restricted to nighttime use. It's difficult for humans or computers to detect a potential threat in this flattened-out landscape.

In the right-hand image the camera's image processor has intelligently remapped the scene to emphasize small temperature differences among the objects. Dark and light areas of the scene are clearly delineated. Details such as the railway's overhead wires, the chain-link fence, and the vehicles on the street are sharp and clear.

In fact, the quality of the image suggests a sharp black and white photo, not the blurry video of past thermal cameras (and of many visible-light cameras under intense sunlight).

Image Correction

Figure 5 provides a good baseline for understanding the degree to which a smart thermal camera can enhance overall brightness, contrast, and detail in a video scene. We will now look at what image processing can accomplish when applied to specific challenges such as environmental conditions and long-range detection.

Piercing the Mist

We have already seen in Figure 3 the degree to which environmental conditions can mask potential threats. That image showed a street during a heavy storm during the day. The rain had brought the temperatures of background objects closer to each other, and very close to the normal temperature of humans. This allowed an intruder to blend into the blurry, poorly defined background.

Figure 6 lets us see that image side-by-side with another image of the same scene, this time taken by a thermal camera with built-in image processing capabilities.

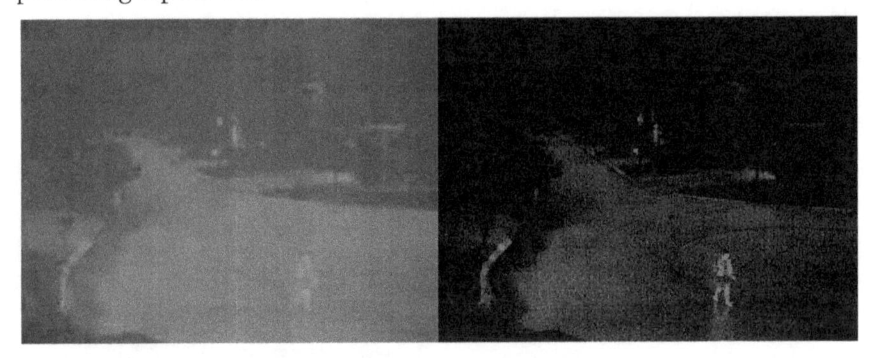

Figure 6: Overcoming the effects of rain and humidity

The original image (shown here on the left) looks washed out, and the person walking through the scene is barely visible. There's a possibility that if this person were an unauthorized intruder entering a restricted area, he or she would go undetected.

There is much more contrast and definition in the right-hand frame, where image processing has intelligently amplified temperature differences among objects. The person stands out clearly, improving the chances of detection, and scene details like the mailbox, the center island of the cul-de-sac, and a street lamp are now visible.

Correcting Thermal Crossover

Until recently a phenomenon known as "thermal crossover" was the Achilles' heel of thermal cameras. Thermal crossover occurs when objects such as buildings and parking lots that absorb heat during the day start to cool after sunset. (This can also occur during the day in certain situations.) At some point, they will "cross over" the 98.6° F mark, matching the normal temperature of human beings.

At this point, at least in theory, a person in a thermal video will appear in the same shade of gray as surrounding objects. He would be indistinguishable from the background.

In fact, however, there are always some temperature differences among objects in a scene, no matter how small. Earlier thermal imagers were not sensitive enough to detect these differences. Even the sensitive cameras now available often cannot overcome thermal crossover because they lack the processing power to clearly differentiate among small variations in temperature.

Today's best cameras use intelligent image processing to enhance the contrast between intruder and background. Figure 7 compares renderings of the same scene from a camera without processing (left) and a camera equipped with image processors.

Figure 7: Targets concealed and revealed during thermal crossover

In this scene someone has walked along hot pavement toward an area shielded from the sun by an overhanging canopy. The temperature in the sheltered area closely matches the temperature of the person, and the person in the left-hand image blends into the background, becoming almost completely invisible.

On the right, the camera's on-board video processing automatically emphasizes the temperature differences in the scene to reveal the potential intruder. It also brings out much more scene detail to provide context for the viewer. Note that you can clearly see the person's head, which is totally invisible in the unprocessed image.

Imaging over Long Distances

Objects that are farther away send less heat energy to the camera than objects that are closer. The air between object and camera further decreases this energy as it travels towards the camera, a problem which worsens as the distance increases. When a thermal camera shoots a scene from long range, many of the finer details of the distant objects will be lost, and backgrounds will lack clarity and appear to be out of focus.

Figure 8 shows how image processing can enhance the energy from distant objects, clarifying details in the video.

Figure 8: Long-range video shot without image processing (left) and with processing (right).

The unprocessed image on the left presents the general shape of the bridge, but not much of its structure or its surroundings. An intruder bent on mischief could easily escape detection by hiding among the beams and wires. On the right, however, the image-enhanced picture presents a highly detailed structure and reveals the tree line, the presence of small boats, and the cloud cover in the background. There's no place to hide.

Removing Naturally Occurring Movement

Detection is based on movement. There is movement everywhere in the outdoor world. In the vast majority of cases it is irrelevant to a security team. Security professionals want their smart thermal video cameras to ignore meaningless motion so they can concentrate on movement that might represent security threats.

While smart thermal cameras are inherently immune to phenomena such as moving headlights and reflections off water that can trigger nuisance alerts in visible-light systems, they can be fooled by other common types of outdoor movement. Image processing prevents this by automatically filtering out the motion of wind-blown foliage, shadows cast by cloud movement, and similar motion artifacts.

Camera Selection Guide: Image Processing

Smart thermal video cameras boast robust, powerful, on-board image processing for stabilization and enhancement. Stabilization enables accurate and reliable motion detection. Enhancement automatically improves clarity, contrast, and detail in any weather and under all lighting conditions. This makes it easier for video analytics and human observers to detect suspicious activity and determine whether it is a real threat. Key features to look for:

- Ability to reveal objects that blend into the background
- Elimination of "white-out" from thermal loading
- Automatic adaptation to low-contrast situations (fog, rain, humidity)
- Detailed images at long distances

Setting Policies

Once an image has been stabilized and corrected for contrast and detail, the smart thermal camera's video analytics engine is ready to detect and track motion, and to determine what threat level it presents, if any. This is where setting detection policies comes in. Here we cover a few basic options to show how the process makes target detection, recognition and identification more accurate and reliable. Chapter 4, The Basics of Video Security System Design, will go into policies in more detail.

Geo-registration and Target Recognition

The prime directive for a thermal security camera is to detect the presence of unauthorized persons or vehicles anywhere in the site and to notify security forces if they are credible threats. To do this, it needs to know where to look for suspicious movement, and what kind of objects and behavior it should be looking for. You don't want it warning the security staff about people at a picnic on the property next door, or detecting and alerting security staff to stray dogs.

Before it can be programmed for target recognition, however, the camera has to go through a process of geo-registration to prepare it for determining its distance from objects in its FOV. Without knowing how far away an object is, particularly in large areas, a camera cannot tell how large it is. Small objects standing close to the camera will appear substantially bigger than a person off in the distance.

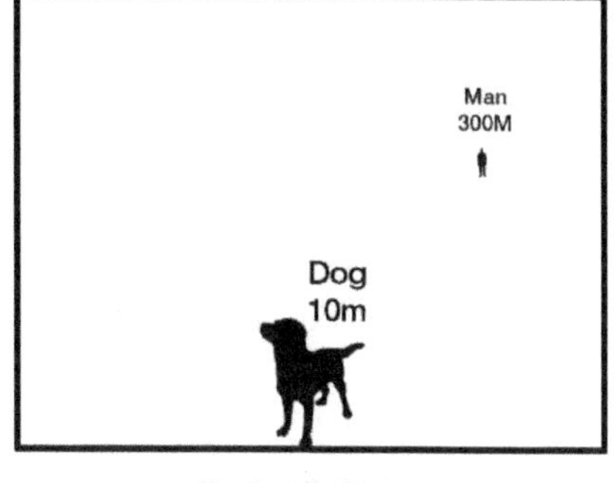

View from the Camera
Note: The Dog at 10M is about 250x larger than the man at 300M

Figure 9: Geo-registration gives cameras depth perception

Figure 9 provides an example. As it shows, a dog at 10 meters from the camera is approximately 250 times larger than a man at 300 meters appears to be. Cameras that lack geo-registration will see the dog as the larger object and send an alert to security, while ignoring the person-sized object in the distance. A camera with geo-registration will be able to adjust its size calculation for distance, determine that the man is actually bigger than the dog, and recognize him as a more credible target because of his size.

Properly geo-registered cameras can detect man-sized objects anywhere within their field of view, whether near or far from the camera, and ignore small animals, blowing trash or debris. They avoid the problem of misreading object sizes, the nuisance alerts it generates, and the security threat it poses in not detecting the man, who is the actual intruder.

Once a camera is geo-registered it's also necessary to establish geographical boundaries – a fairly simple procedure. Then you can program the camera to restrict its alerts to certain sizes of target. On a site that includes roads and water features, for example, you might program parameters such as these:

Target	Size	Speed
Human being	1.5 - 2.2m × .05 - 1.0m	1-15 Km/h
Vehicle, front view	2.0 - 2.6m × 3.0 - 3.5m	5-100 Km/h
Inflatable boat, front view	3.0 – 3.5m × 4.0 – 4.5m	3-30 Km/h

And so on. These rules will maximize the likelihood that the camera will detect objects of special interest to security personnel and alert staff to their presence, while ignoring the squirrels and deer on the property. Smart cameras make assigning these rules intuitive and easy.

We will walk through the actual steps for geo-registering and programming a smart thermal camera in Chapter 4.

GPS and PTZs

Smart thermal cameras that are geo-registered to the scene also have the ability to provide global positioning satellite (GPS) location information to take the detection and recognition process to the next level: teaming with a visible-light camera to obtain positive identification of intruders. This feature will eliminate unnecessary reconnaissance trips by the security staff.

In this scenario the smart thermal camera sends GPS coordinates for the target, directing a pan-tilt-zoom (PTZ) visible-light camera to zoom in on that location. High-definition PTZs can provide highly detailed images of intruders for identification purposes.

Built-in GPS functions bring another advantage to a security operation: mapping the location of the intruder onto a site map to improve response time. When the staff judges that it must investigate in person, GPS-equipped thermal video security systems will give them the precise coordinates of the apparent breach so they can get there by the shortest and most direct route.

Camera Selection Guide: Policies

Setting policies to customize automatic detection and recognition functions to your needs can reduce nuisance alerts by a factor of 100, and eliminate a major source of fatigue for the security staff. Look for cameras that

- Are easy to program
- Incorporate geo-registration
- Use GPS functions to guide PTZ cameras, and provide target coordinates to security staff

The following chapter will examine the cost – and cost savings – of installing a smart thermal video security system.

3

How Smart Thermal Video Cuts Security System Costs

Conventional wisdom says you can't put a price on security. Unfortunately the real world doesn't always work that way. Every outdoor security system has a price, and anyone who plans, purchases, or uses such a system has to find a price that fits the budget.

Today's thermal video security technology makes it much easier to balance the desired level of security protection against the available budget. The best systems are accurate and reliable. And remarkably, they are also less expensive than other, less effective systems.

In fact, they deliver their more reliable, more accurate protection at up to 50% less in total cost when compared to systems built around visible-light technology, which require many more cameras. Fewer cameras equals less infrastructure, and in all but a few cases infrastructure (camera mounts and installation, cabling, and construction) is the biggest driver of expense.

This chapter outlines the factors that affect the total cost of a video security system, and shows how a high-performance smart thermal system saves money. It does not price out specific equipment or individual systems. That would be impractical, considering how greatly outdoor sites vary in size and configuration. Its intent is to provide security professionals with rough guidelines for

- Evaluating the competing technologies
- Planning cost-effective systems
- Controlling infrastructure and installation costs
- Reducing wasteful staff chores

We start with a case study showing how a smart thermal video security system can reduce installation and operational costs when compared to a visible-light video surveillance system.

Thermal Video vs Visible-Light: Budget and Performance

As noted in Chapter 2, smart thermal video security cameras with built-in image processing will detect motion and recognize a potential threat at greater distances and over wider areas than visible-light cameras with analytics. The latest smart thermal cameras cover areas the size of a football field (Figure 10), more than ten times (10X) what the visible-light competition or even an infrared camera with external analytics can handle.

The obvious security benefit of broader coverage by a security system happily translates into significant cost reductions as well.

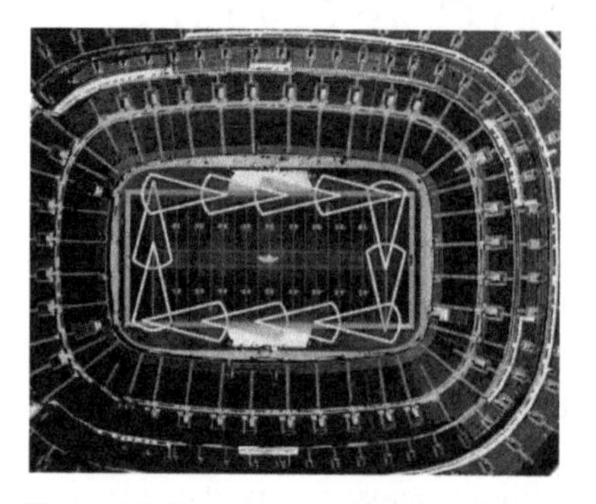

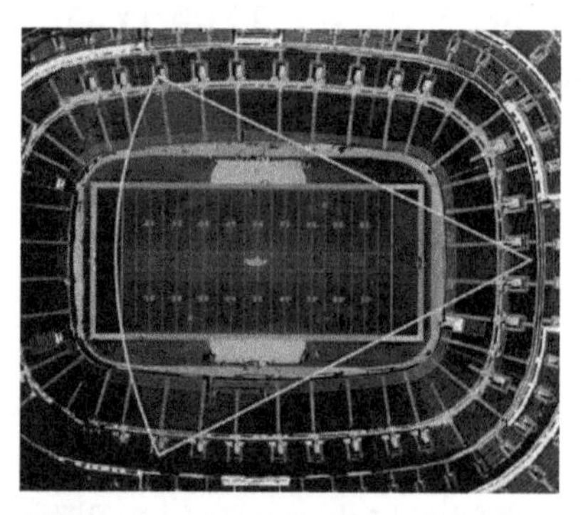

Figure 10: Smart thermal camera (image on right) covers an entire football field, more than 10X the area of the alternative technology (left image)

Savings start with equipment costs. Since the cameras cover more than 10X the area, the number of cameras required for full coverage drops significantly.

As noted above, when fewer cameras are needed, the impact on system cost is profound. With fewer cameras the site needs fewer camera mounting poles. That triggers an avalanche of savings in construction costs. There's a drastic reduction in the need for power lines and wired network connections. There's less trenching or stringing of overhead wires.

Figure 11: Smart thermal cameras are well suited for vandalism challenges at rail yards

Vandalism Solution

For a better sense of how this works in actual practice, consider the case of a major passenger rail company in Italy. The company deployed smart thermal cameras to provide rail yard security and prevent vandalism at one of the largest railcar maintenance facilities in the country. Its experience is typical of what security professionals can expect in terms of decreased cost and increased ROI when they choose longer-range and more accurate outdoor security cameras over limited-range alternatives.

The rail organization serves more than 650,000 people every day and nearly 180 million passengers every year, delivering crucial transportation services for Milan's financial district and the surrounding region. The company's expansive rail yard stretches over two kilometers, encompassing an area equivalent to sixty football fields, where millions of dollars' worth of trains are stored and serviced.

Located along a busy public highway, the rail yard was experiencing a difficult security challenge. Intruders were able to gain access to the yard undetected at night and cause severe damage to rail cars stored at the facility. It was a huge, expensive problem. Restoring an individual rail car costs many thousands of dollars, and it has to be removed from circulation to be repaired, which incurs lost revenue as well.

The ongoing vandalism occurred even though the rail company had guards patrolling the large rail yard. Guards in that region of Europe can cost as much as $20,000/month, so the company was spending large sums on security without getting any closer to its goal of detecting intruders along the large perimeter of its yard at all hours of day and night, even in complete darkness.

When a newly purchased line of rail cars was badly damaged by vandals, the organization recognized the need for a more effective solution. Initially they considered using visible-light surveillance cameras. They turned to their security integrator, who explained that surveillance cameras would still require active monitoring by security personnel, and would be ineffective at night when many of the intrusions were taking place.

Instead the security integrator recommended smart thermal cameras to detect intruders in real time. The system was well suited for the large perimeter and would detect with accuracy regardless of lighting and weather. The integrator was confident the system would meet the organization's security needs at much lower infrastructure and operating costs.

> "The company was spending thousands of dollars on guards and was not close to the goal of detecting intruders day and night."

Accordingly twelve long-range smart thermal cameras were installed to provide automated awareness along the two kilometer perimeter. When a camera detects an intrusion, onsite security personnel receive an alarm along with the intrusion's location pinpointed onto a topology map of the facility. For real-time visualization, high definition PTZ cameras are automatically steered to zoom in on and follow the intruders, allowing security guards to quickly assess the situation. Once an alarm has been verified, local police authorities are notified to respond and intervene.

According to the integrator, the solution has provided excellent outdoor detection with very few nuisance alarms. "Maintenance costs have been very low, and the cameras have stood up well to the elements. On the installation side, it's a very easy system to deploy," they stated.

Shortly after the solution was put into place, the technology detected several trespassers entering the yard. The system alerted rail security to intervene and local authorities were able to apprehend the intruders. The organization is expanding the system to other maintenance facilities throughout the region.

Factors for Success

The rail yard's successful implementation highlights a fundamental change in smart thermal video security: it is not only technologically superior to other approaches – it is now cost effective as well.

Thermal cameras once had a reputation for being very expensive. But like most other electronic devices they are beneficiaries of advances in design and manufacturing that boost performance while reducing cost. Today they make perfect economic sense as the foundation of an advanced video security system.

That doesn't mean they cost less than their visible-light counterparts. Thermal cameras, especially one of the smart models that are revolutionizing outdoor site security, remain somewhat more expensive. But the gap between the two technologies is much smaller. And when one smart thermal video camera can replace eight to 10 visible-light cameras with analytics, the overall savings far outweigh the modest per-camera price premium.

Factor in the lower outlays for infrastructure and installation that naturally follow when you reduce the camera count, and a smart thermal video security system becomes the low-cost choice. Let's look at one example of how such a system can impact the budget for a new or upgraded system.

Higher Performance, Lower Cost

Video security and surveillance systems are most commonly configured to protect the perimeter of an outdoor site. However, another strategy is becoming increasingly common: area or asset coverage.

Figure 10 above illustrates this approach: a single smart thermal video camera protects an entire football field, including its perimeter, while it takes 10 visible-light cameras with external analytics to cover only the perimeter.

Table 1 quantifies the cost savings that accrue from using this strategy. In this example a single smart thermal camera monitors a 48,000 ft^2 buffer zone. Ten competing cameras with external analytics are needed to cover the same area. The smart thermal camera is 14X the price of one of the competing cameras, yet the total cost of the smart thermal video system is *less than one-fourth* that of the other system.

Outdoor Security Task	Cost of smart thermal cameras and infrastructure	Cost of other cameras and infrastructure
Monitor buffer zone of 48,000 square feet	1 camera: 9,995	10 visible cameras @ $700: Total $7000 3 Analytic Appliances @ $2000: Total $6000
Infrastructure (Poles, Power, Communications)	1 Pole : $5,000	10 Poles: $50,000
Total system cost	**$14,950**	**$63,000**

Table 1: Greater coverage by a smart thermal video camera results in significantly lower system cost

This startling difference comes from substituting long-range, wide-area coverage smart thermal cameras for limited-range, narrower-coverage devices.

There is a wide range of coverage patterns available in the latest smart thermal camera models (see Table 2 in the next chapter for a list). This allows system planners to design sites with the minimum number of cameras necessary. Because of this, the average smart thermal video system will have a total cost at least 50% below what can be achieved with standard cameras, and often much lower.

The economics are even more compelling when ongoing costs are factored into the budget. Since a smart thermal video system is much less likely to trigger nuisance alerts, security staff will not be forced to waste time and money checking on false alarms. Fewer trips translate into reduced operating expenses.

Reducing the Cost of Perimeter Protection

In many high-risk situations it is desirable to create zones of protection that extend outside the fence-line. Such buffer zones give security staff more time to respond before intruders breach the perimeter and get to the site's assets.

Typically a fenced perimeter is protected by a "blind" sensor such as fiber on the fence, infrared beam, or passive infrared sensors (PIRs). The sensor is supposed to detect activity that might pose a threat to the site before the potential intruder can breach the fence. It is augmented by a camera to help security guards determine the cause of any alerts it generates.

The sensor approach usually acts as a supplement to a site's video security system. But the cost of deploying and maintaining two separate systems – sensor and video – can quickly escalate.

Nor is it money well spent. A "blind" sensor has limitations. Like the video system it is intended to supplement, a blind sensor is prone to a very high rate of nuisance alerts. Deer brushing against the fence can set off its alarms.

In addition, potential intruders must touch the fence (or be in very close proximity) before they are detected. Clever trespassers, both animal and human, can find ways to circumvent that protection by bridging over or tunneling under the sensing system. These "blind" sensors add substantial cost to a security system, but very little security.

If a security manager needs an early warning function, he or she is better advised to implement volumetric coverage of the fence, as illustrated in Figure 12, to provide a "buffer zone" outside the fence. Not only does this eliminate the need for an expensive "blind" sensor system, it is also a best practice for outdoor security.

Smart thermal cameras excel in the role of volumetric sensors. They detect people and objects beyond the fence line of an outdoor site, triggering alarms when intruders approach the facility. Security personnel get ample warning of the threat and have more time to intervene.

Figure 12 compares a smart thermal video security system in volumetric mode (left) with a sensor system (right).

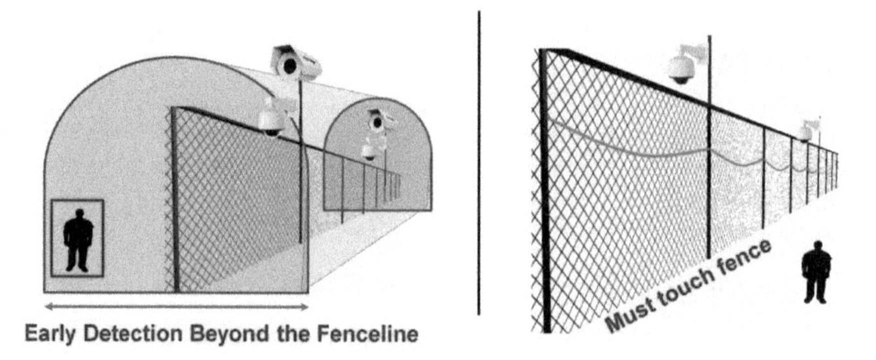

Figure 12: Early detection beyond a fence

It's immediately apparent that their wider coverage gives thermal cameras an advantage in anticipating the approach of an intruder.

There are some high-risk situations where a security manager might decide that a site needs a second layer of protection, and choose to deploy a fence-mounted sensor system as a supplement to a smart thermal video security system. Even if he or she takes this approach, the resulting hybrid system will have a lower overall cost than the system on the right in Figure 12, because the superior coverage of a smart thermal video security installation will still require fewer cameras.

Alarm Policy-Driven Video Security

Structuring a security system often becomes a tug-of-war between comprehensive protection and budget constraints. Up to now we've explored how smart thermal video security systems go a long way toward reducing the tension by

- using fewer cameras than a conventional system, and
- requiring less infrastructure and thus limiting the cost of installation
- greatly increasing security and situational awareness

And there is another factor that helps hold down costs: establishing appropriate detection and response (alarm) policies. Although this does not affect initial cost, its impact in lowering the ongoing operating costs of a security program is substantial.

Smart thermal video cameras are designed to detect objects that violate video analytic rules. They place a red box on violations, send alerts over the security network, and track the potential intruders through their fields of view.

The result is better targeting of truly suspicious movement. Security staff will not be forced to respond to a large number of nuisance alarms from objects that do not threaten the site. The downstream savings in lost man-hours and wear and tear of response vehicles can be considerable.

The next chapter will include a section on setting alarm policies, within the context of how to select equipment and design a smart thermal video security system.

Part Two

Creating a Smart Thermal Video Security System

4

The Basics of Video Security System Design

Security systems have four key objectives:

- **Deter**. Discourage potential intruders from entering a site by the use of signage, audible sirens and alarms, loudspeaker announcements
- **Detect**. Find actual intruders and track them through the site; alert the guard station to the trespass; display imagery of the potential threat; and direct security forces to the affected area
- **Delay**. Use a fence, a roadblock, water, hedges, or other landscape features to slow down intruders on their way through a site
- **Deny**. Keep intruders away from site assets with electrification of the fence around the perimeter and staff response

Some experts expand on these concepts. For example, the North American Energy Reliability Corporation (NERC) suggests a slightly different "systems approach" to assure the physical security of vulnerable electrical substations.

In its Critical Infrastructure Protection standard CIP-014 (referred to in Chapter 1), NERC outlines a total of six functions. The first three are the same as above: Deter, Detect, and Delay. But in place of Deny, NERC specifies three more specific steps: Assess, Communicate, and Respond.

Creating an outdoor security system around smart thermal video technology allows us to implement or control several of these functions with a single intelligent device. Table 2 matches the capabilities of smart thermal cameras against the requirements of a total security system. (Delay applies exclusively to physical barriers, so it is not included.)

How Smart Thermal Cameras Detect, Deter, Assess, Communicate and Respond		
NERC CIP 014	**Smart Thermal Cameras**	**How They Help Meet NERC CIP-014**
DETECT	Accurately detects intruders at all times, in all conditions	Detects unauthorized intrusions before incidents escalate
	Detects intruders, ignores environmental movement	Reliable alerts provide security accountability
	Volumetric Coverage	Defense in Depth detects beyond the fence line, provides early warning and more time to respond
	Buffer Zone Protection	Layers of protection over vulnerable access points throughout the site
DETER	Audio Alarm Warning	Notifies detected intruders to leave the site
ASSESS	Visual Situational Awareness	Real-time site map displays camera field of views and target locations
	Red Box Placed Around Intruder	Highlights detected target for fast response
	Auto Steering of PTZ	Pans and zooms PTZs onto target for immediate classification
COMMUNICATE	Alarms, video, and target info about threats sent to central reporting locations	Real-time intelligence to process, analyze, evaluate, and respond to threats
RESPOND	Intruder location projected onto sitemap of facility	Responders know the precise location of each intrusion in real time

Table 2: Building a security system around smart thermal video technology

From this comparison we can see that a properly designed smart thermal video security system can supplement, control, or actually replace other measures often used to implement essential security functions. For example:

- Volumetric video detection on a site's perimeter detects beyond the fence line
- Map-based displays precisely locate an intrusion, making response easier
- PTZ steering allows close-up display of a detected intrusion for more reliable assessment of the level of threat.

We will now look at what it takes to create such a system.

Equipment Requirements

A successful smart thermal video security system starts with putting the right hardware in the appropriate places. There are three major decision points in this process:

- Choosing the right video cameras for the site
- Locating them for total coverage with no blind spots
- Creating the necessary infrastructure to support the system

Selecting the Cameras

There are two fundamental questions that determine what equipment to choose for your site. What kind of cameras should you use? And how many do you need for full coverage?

By now it's clear that smart thermal video cameras are the foundation of an accurate security system. But most sites, especially larger ones, should have a combination of thermal video cameras and visible light units with pan/tilt/zoom (PTZ) capabilities, or IP dome cameras. The thermal cameras will provide the basic security functions, including detection, initial assessment, and tracking of potential intruder "targets." The PTZ cameras will support these functions by showing close-up views of any detected targets so security staff can identify the exact nature of a threat.

The criteria for selecting thermal cameras are somewhat different from those for visible-light cameras.

Thermal Camera Detection Ratings

Reliable site security requires seamless video coverage of the perimeter. The most important factors in achieving this are the reach of the cameras and their field of view (FOV, or lateral coverage).

The reach of smart thermal cameras is rated according to their automated detection range. This is the maximum distance at which a thermal camera can reliably detect a human being in motion.

There are two ways to measure the automated detection range of a camera: with a subject walking across the camera's field of view, or with the same person walking "inbound" or directly toward the camera. A person walking toward a camera does not appear to be moving much compared with a person walking across the field of view. The camera sees mostly leg motion, which makes it harder to detect the target.

Figure 13 below illustrates the inbound measurement procedure. This is the best way to rate a security camera, because it gives you a real-world, worst-case figure for reliable detection. You are assured that the camera will always detect potential intruders within that range.

Figure 13: Determining a camera's true detection range

When examining the specifications of cameras, make sure the manufacturer specifies "inbound detection ranges." Unless the camera specification explicitly states "Inbound Detection," assume it is crossfield, and test its inbound range yourself.

Also consider the environmental factors that affect inbound detection. Make sure the specifications for any camera under consideration include the distance at which a pedestrian can reliably be detected under conditions that reduce image contrast, such as rain, snow, and fog.

These precautions will help you design a system without security gaps.

Thermal Camera Site Coverage

For total security on an outdoor site you obviously have to find cameras that can detect suspicious movement from one end of the site to the other. Since thermal cameras are generally not equipped with variable zoom lenses, you must select those with fixed detection ranges that match the dimensions of your site.

- Longer detection ranges make better economic sense for very large sites, because they reduce the number of cameras needed for full coverage. At this writing the maximum available detection range is approximately 600m (1,965 ft).
- For distances too long to be covered by a single camera, use the fewest number of cameras that will do the job. The longer their automated detection range, the fewer cameras you will need.
- On smaller sites, or for blind spots within a larger site, choose cameras with the minimum detection range for the dimensions being covered. Detection ranges as short as 35m (115 ft) are available.
- Large areas and buffer zones benefit from using a 640 x 480 camera which covers about 4 times the area of a 320 x 240 camera with the same field of view.

The other camera characteristic to keep in mind when determining coverage is FOV, which is measured in degrees of coverage. It's a fact of optics that the longer a camera's reach (detection range in thermal cameras), the narrower its FOV. Table 3 shows some typical thermal camera coverage specifications.

320 x 240 Smart Camera			**640 x 480 Smart Camera**		
Automated Detection Range of a Person	Coverage Area	FOV	Automated Detection Range of a Person	Coverage Area	FOV
35m (115ft)	949m² (10,216 ft²)	90°	75m (246ft)	4,408m² (47,453 ft²)	90°
60m (197ft)	1,851m² (19,919 ft²)	60°	100m (328ft)	6,084m (65,483 ft²)	70°
95m (312ft)	3,244m² (34,921 ft)²	42°	145m (476ft)	9,089m² (97,835 ft²)	50°
160m (525ft)	5,248m² (56,491 ft²)	24°	195m (640ft)	12,373m² (133,184 ft²)	37.5°
242m (794ft)	7,992m² (86,026 ft)²	16°	300m (984ft)	19,363m² (208,425 ft²)	25°
313m (1,027ft)	10,357m² (111,482 ft)²	12.4°	410m (1,345ft)	25,658m² (276,177 ft²)	17.6°
430m (1,410ft)	14,213m² (152,986 ft)²	9°	595m (1,952ft)	38,071m² (409,790 ft²)	12.4°
600m (1,969ft)	18,979m² (204,298 ft)²	6.2°			

Table 3: Smart Camera Coverage Specifications

FOV combined with detection range creates a cone-shaped detection zone with the pointed end of the cone closer to the camera. Smart cameras with wide detection zones (640 x 480) are ideal for volumetric coverage of an asset as well as establishing a buffer zone along perimeters (see the Detect row in Table 3).

Geo-Registration

Later in this chapter we will cover the various programmable functions available in many smart thermal video cameras. Here, however, we will take a brief look at geo-registration, the function that enables thermal cameras to provide precise coordinates for the location of any intrusion. This will help us understand how thermal cameras can instruct PTZ visible-light cameras to focus on that exact spot.

As mentioned in Chapter 2, some smart thermal cameras have a geo-registration function. This allows the mapping of all objects in a camera's FOV to GPS coordinates. The camera's software uses these references to determine the true size and location of every object it sees. It won't favor a dog over an invading human just because the dog is closer to the camera and takes up more of the frame. It can figure out that the human, which looks smaller because it is farther away, is actually bigger, and is a potential threat. If the camera detects such a threat it will alert the security force, providing accurate GPS coordinates for the object so the staff can investigate quickly.

Add the right interface and the smart thermal camera can go one step further. Using the GPS information, it can steer a PTZ camera to focus on the location of the intrusion.

PTZ Functionality

PTZ cameras are often used to scan outdoor areas. Affordable PTZs are visible-light devices, so when they are used by themselves they provide only passive surveillance. Match them with smart thermal video cameras, however, and they become an active and valuable adjunct to an outdoor site security system.

As the name implies, these cameras can pan 360^0 to cover all corners of a site, tilt up and down to focus on different areas within the site, and zoom in on an area or target of interest, giving security staff a clearer view. The zoom capability gives them an advantage over most thermal cameras in displaying detailed imagery of a potential threat.

But while PTZ cameras can be useful for validating the nature of an alert, there's almost no chance a PTZ will be looking in the right place when an intrusion occurs. Its narrow FOV relative to the wide areas under surveillance almost guarantees that events will go undetected, and trying to steer PTZs over a large area manually is like searching for a needle in a haystack.

A PTZ camera steered by GPS coordinates from a smart thermal camera is a different story. It will pinpoint the location of an alert, zoom in on it, and follow the activity in real time. The resulting imagery gives security operators what they need to quickly assess the nature of the alarm and react appropriately, as shown in Figure 14.

Figure 14: Long-range smart thermal cameras for detection (left) and auto-zoom PTZs for assessment (right).

There are many PTZ cameras available today. One of the leading makers offers models with zoom ratios from 12x to 28x. Select the one that can fill the display screen with the information you would like to capture (a face, a license plate number) when the object is at your site's perimeter and the camera is zoomed in to maximum telephoto.

Avoiding Blind Spots

A camera's field of view doesn't begin where it's mounted. It can only detect at a measurable distance in front of it. The area in between is the camera's blind spot, and your security design has to take it into account, or someone will be able to walk right under a camera undetected.

Following one simple guideline will make sure this does not happen: *The view of each camera should cover the next camera's blind spot.*

Take, for example, sites with one camera per side. Some outdoor surveillance designs will narrow a security camera's field of view to increase its detection distance (Figure 15) in an effort to decrease costs by reducing the number of cameras needed.

Figure 15: Narrow FOV creates blind spots

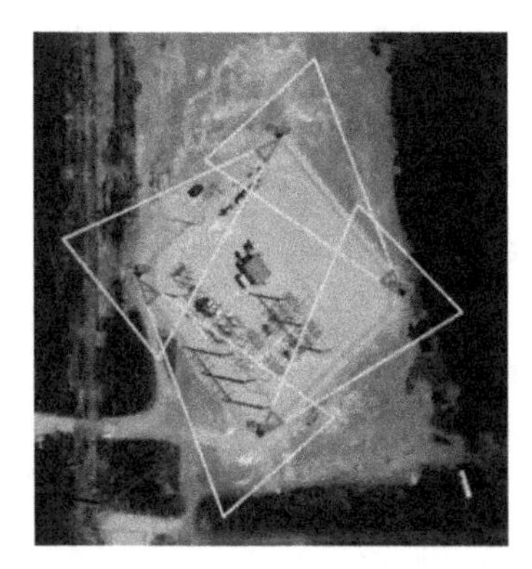

Figure 16: Wider FOV provides total coverage

This is not necessarily a bad concept, but it's important to understand that doing so also makes the blind spot under each camera larger. You must then add blind spot coverage, sometimes doubling the number of cameras required. It's a strategy that all too often backfires in economic terms.

Figure 16 shows how using cameras with a wider FOV solves the problem, making "blind spot cameras" unnecessary and saving money in the process.

The principle of having one smart thermal camera cover the blind spot of the next one also applies when one side of a site perimeter is so long that you need more than one camera to cover it. Figure 17 demonstrates how the blind spot of one security camera can be covered by the one behind it.

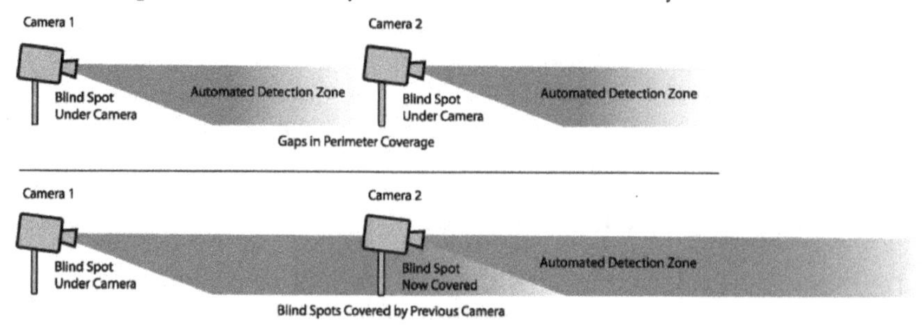

Figure 17: Addressing camera blind spots

In this case the cameras are mounted at twenty feet off the ground with a seven degree horizontal field of view. In the top design, the coverage range of "Camera 1" stops near the base of the pole of "Camera 2," leaving approximately 60 meters of unprotected area where intruders can enter undetected.

The secure approach is shown in the bottom design. Here, a longer-range camera is used to extend the automated detection zone of "Camera 1" to cover the blind spot under "Camera 2."

Choosing the right equipment and making sure it can cover the entire site without gaps in coverage is crucial to the success of a security system design. We now turn to a more practical matter: putting the system in place.

Designing the Support Infrastructure

Outdoor security projects encompass many costs. These include engineering firms to design plans for construction and trenching, contractors to install electrical and communication wiring, and on-site provisioning. In just about all cases these expenses add up to more than the cost of the cameras.

In Chapter 3 we saw that building your system on smart thermal video technology and selecting automated detection ranges and FOVs appropriate to your site will reduce infrastructure costs by as much as 50%. You need less equipment because smart thermal video cameras cover more area than visible-light units. That translates into a reduced need for installation and support. And the longer or wider the range of these cameras, the fewer you will need.

But you can also employ other, less obvious strategies related to camera enclosures and mounting methods that will further reduce infrastructure requirements.

Packaging, Poles, and Mounts

How a camera is protected from the elements is a significant factor in determining the reliability of a security system, and that has a long-term impact on your budget. Replacing a camera that's malfunctioning because of exposure to the elements can be a costly proposition, particularly if it's sitting at the top of a 40-foot pole.

Many so-called outdoor video surveillance cameras are really indoor cameras with an add-on protective enclosure to keep out rain. They are susceptible to extreme temperatures, sand, dust, and humidity. Even moderate changes in outdoor temperatures can cause expansion and contraction in the housing, allowing grit, dust, or humidity to enter and affect the electronics and optics. This exposure degrades the camera's ability to accurately detect targets and shortens its service life.

Figure 18: Security camera in a NEMA enclosure

Outdoor cameras must be designed from the outset to withstand changing temperatures and extreme environmental conditions. To protect your equipment investment and avoid the ancillary costs of replacement, choose cameras in nitrogen-pressurized housings, sealed to handle temperature changes and keep out humidity, snow or even blowing sand.

Look for a NEMA 4X or IP66 (or higher) rating in the camera's specifications. This identifies a camera that will operate reliably even in such environments as the -58^{0}F cold of the Canada Oil Sands or the 158^{0}F heat of a Middle East desert. Figure 18 shows a security camera in a NEMA enclosure mounted on an existing building (see below for advantages of building mounts).

Mounting Locations

Large sites usually require remote cameras positioned far from the security operations center. These are often mounted high off the ground on poles to get a clear view of the perimeter and to discourage tampering.

What kind of pole you use can affect camera performance. Concrete, aluminum, or steel poles are preferred because of their stability and their resistance to weather. Avoid telephone/utility and standard wooden surveillance poles. Not only do wooden poles sway in the wind, the wood will shrink, causing detection zones to shift unpredictably.

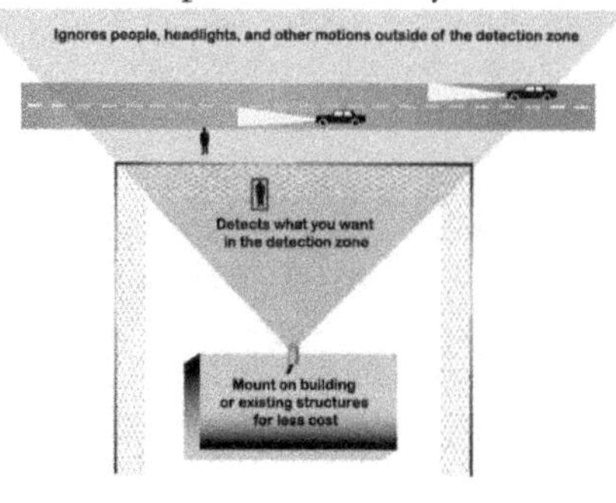

Figure 19: Geo-registered camera on building covers site perimeter

Existing buildings and other structures also provide good mounting locations for cameras, as seen in Figure 18 above. The mounts are easier and much less expensive to install than poles.

In addition, many structures will already have power and communications wiring in place, eliminating the cost and trouble of running new cables.

As Figure 19 illustrates, smart thermal video cameras with geo-registration have the flexibility to take advantage of building mounting by programming the camera's coverage zone according to size and location. Simply aim the camera directly at a fence line, which is the outer border of the zone. The camera correctly detects intrusions the instant they enter the secure area while ignoring objects outside the detection zone.

Wired or Wireless

You can eliminate many infrastructure implementation costs by avoiding the need to run power and communication lines to pole-mounted cameras.

Choose power-efficient cameras and you may be able to run them on batteries, with a solar panel to recharge the batteries during the day. Fortunately, today's smart thermal cameras draw far less power than they did just a few years ago. The latest model from one of the industry's leading suppliers delivers 4X the processing capability using only 40% of the power of its immediate predecessor.

Likewise, wireless networking removes the need to run communications lines. We will revisit these concepts in Chapter 6.

Implementing Functions

As detailed in earlier chapters, smart thermal cameras feature powerful built-in processing that analyzes video to detect objects that violate a site's alarm policies. A video processing board sits inside the camera housing to both digitize and analyze video in real time, while stabilizing the video to ensure a clear, stable image and enable even small objects to be detected.

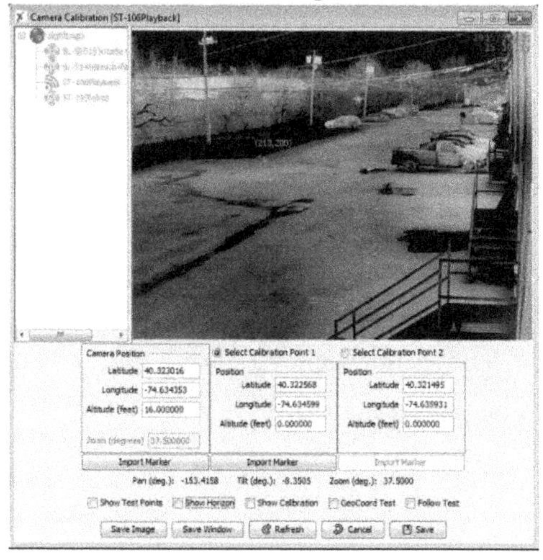

Figure 20: Camera location and calibration

Each system is highly configurable. You can adapt it to the exact requirements of your individual site. This includes setting alarm policies that specify when and where alarms are generated, and defining the types of objects that can trigger alarms. You can even adjust video bandwidth to stay within the site's network capacity.

That configurability lets smart thermal video cameras achieve many of the objectives outlined by well-established guidelines, such as NERC's CIP 014 document. Of course the security operator has to define and activate detection parameters relevant to the site and assets being secured.

Fortunately this is not difficult. In modern cameras it is a simple matter of configuring on-screen menus and either choosing among options or providing information.

Let's look at a typical user interface to show how a few of the more important camera functions can be adapted to a specific site.

Establishing Coverage

When a smart thermal camera is first installed, its actual physical location has to be entered into the system, and then the camera must be calibrated. This establishes the area of coverage for that particular camera. It also allows geo-registration, which enables the camera to determine the location and size of objects they detect and to steer PTZ cameras for close-up views.

Figure 20 shows how this is done. The client software shows imagery from the camera along with a menu for setting camera position. The user enters the camera's latitude, longitude, and altitude (meaning height from the ground) – into the "Camera Position" boxes on the left by viewing the camera's physical location on Google Earth and clicking on it. The values can also be entered manually.

Once the camera's position is registered, it's an equally simple matter to calibrate it for geo-registration purposes. This allows the camera to determine the true size and location of every object it sees. It also gives it the ability to direct security operatives to the spot where an intrusion has occurred, and to steer a PTZ camera to that location for a close-up view.

Geo-registered calibration is done just like camera position, with one additional step. The user first clicks on a point in the image. He or she then enters its location in the boxes on the right side of the menu exactly as before. This operation must be done for at least two different points in the image. The camera's processor can now determine exactly how far away each object is, and calculate its actual size.

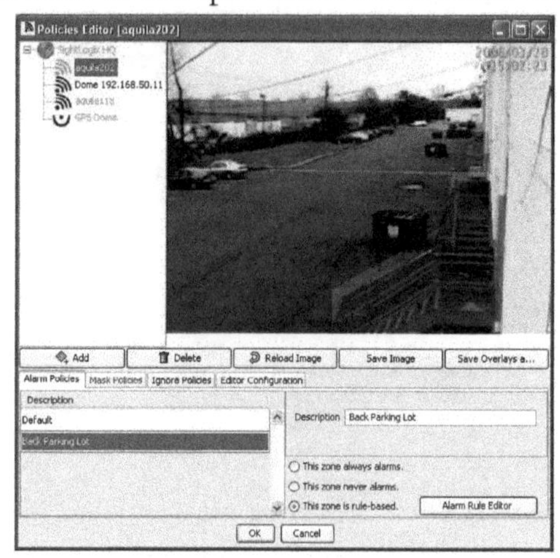

Figure 21: Zone policy menu

Setting Alarm Policies

You control alarm generation by setting up zones and applying alarm policies to them.

By default, any object moving within an alarm zone generates an alarm; however, you can set the system to be more selective as to when alarms occur, even in an alarm zone. For example, you may not want alarms generated during working hours for objects that move within a work zone; or you may want only certain objects—differentiated by speed, size, or heading—to generate alarms. There are many other parameters you can set to define what kind of movement constitutes a threat or an actual intrusion.

Types of Zones

A zone is any part of the area covered by a camera that you define. You can then specify whether to generate alarms or even track moving objects in this zone.

One leading smart thermal camera system supports three types of zones: alarm zones where moving objects can generate alarms; mask zones where all movement is ignored; and ignore zones where objects originating in the zone are not tracked (though currently detected objects continue to be tracked.) Figure 22 represents the menu screen for setting those policies.

In this case the area in the top right of image has been designated an ignore zone. Objects originating here will not be tracked. However, objects already being tracked that move into the ignore zone will continue to be detected.

The large area in the lower left is an alarm zone, and all moving objects there will be tracked if they satisfy the policies set for that zone.

The mask zone in the lower right does not have motion tracking, while objects in the undefined area in the upper left are tracked but do not generate alarms.

Various tabs in this and subordinate menus allow the operator to adjust these parameters. Flexibility is built into the system, allowing you to map out areas that you want to protect, and determine the degree of protection.

Alarm Rules

Every outdoor security system should first and foremost help the security force work more efficiently and effectively. The key to achieving that goal is to alert them to every credible threat while reducing false alarms to the vanishing point.

You can program highly specific rules and policies about what to track and what to ignore into smart thermal video cameras. This gives the cameras the ability to discriminate very accurately between potential breaches and objects that pose little or no threat. Figure 22 shows a menu for setting these rules.

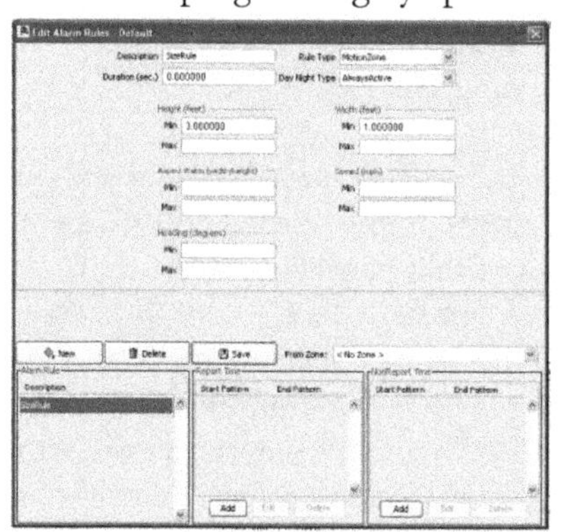

Figure 22: Alarm rules menu

The first step in setting any type of rule – time range, detection type, object attribute – is very straightforward. You first create a rule set as follows:

- Click New and assign a name to the rule set (you'll be renaming it from a default name)
- Select the alarm rule at bottom left
- Define one or more rules
- Click Save

You can then designate a variety of rules that you want applied to the detection and tracking of moving objects. The parameters include

- Designating a zone for detection behavior
- Specifying a from-zone (only objects that enter from a specific zone will be alarmed)
- Setting times when the rules are in effect
- Specifying target attributes: height, width, direction of movement,

speed, aspect ratio, etc.

The parameters you select for detecting and tracking objects will determine the accuracy of the alerts the system will deliver and the volume of policy-based alarms it will generate.

Fortunately, as you can see, it's relatively easy to adjust the rules for maximum efficacy.

To review, the rules can include such parameters as:

Area of interest	Mask out areas not of interest
Minimum object width	Minimum object height
Maximum object width	Maximum object height
Minimum duration of time object must be in defined area	From – To zones
Minimum object speed	Maximum object speed
Object direction, if applicable	Time intervals, for any rule
Multiple zones, if necessary, and each zone has its own set of rules	Tracked objects (a yellow box around every object that the analytics see)

Once the relevant parameters are set, the smart thermal camera will only alert on objects that constitute threats under these rules. The camera has been customized to the security concerns of that particular site.

In this chapter we have tried to present some of the major considerations to keep in mind when designing a smart thermal video security system for outdoor sites, along with a description of the tools available to help in this effort. In Chapter 5 you will see sample sites that users of smart thermal video systems have created to protect their sites and the assets they contain.

5

Model Site Designs

Smart thermal video security systems safeguard thousands of outdoor sites and their crucial assets, protecting them from unwanted intruders. They are on duty at airports and railroads, prisons and government properties, utility infrastructure sites, data centers and commercial buildings, bridges, and other types of facilities around the world.

This chapter presents eight sample designs for smart thermal video security systems on outdoor sites. Each design is accompanied by a brief description of the site, the security problems it presented, and how the system solved these problems in a cost-effective way.

These are designs for actual outdoor sites, not theoretical mockups. Some of them have already been installed; others are in the planning stage. Their applications include transportation hubs, corrections facilities, and commercial buildings. They can serve as models to be adapted for other, similar sites.

Perimeter Security and Area Coverage

Five of the eight designs are representative of the traditional perimeter protection form of security: watching the borders of a site to detect trespassers immediately. The other three depict a newer approach: "asset" or "area" protection, for sites where it makes more sense to protect the assets on the site than to try to guard its entire perimeter. Smart thermal video cameras offer both long-range detection potential and volumetric coverage capability, making them effective in both approaches.

Alert Monitoring Options

The text also specifies how each site is monitored. The standard approach is to have alerts handled by an on-site security operations center (SOC). Recently, however, security managers have been choosing to contract with central station (CS) service companies for off-site monitoring.

Online Design Tool

All eight designs were created using the free, patented SightSurvey online design tool from SightLogix. You are welcome to log into SightSurvey to create a security system specifically tailored to your site. SightSurvey is available at http://www.sightlogix.com/sightsurvey-tool/.

Application: Critical Infrastructure

Electrical Substation

Coverage: Perimeter
Monitoring: On-site Security Operations Center

Smart thermal video cameras create an accurate, reliable security zone around this large site in spite of its irregularly shaped perimeter. Special care was taken to limit the number of cameras and mounting poles in the installation.

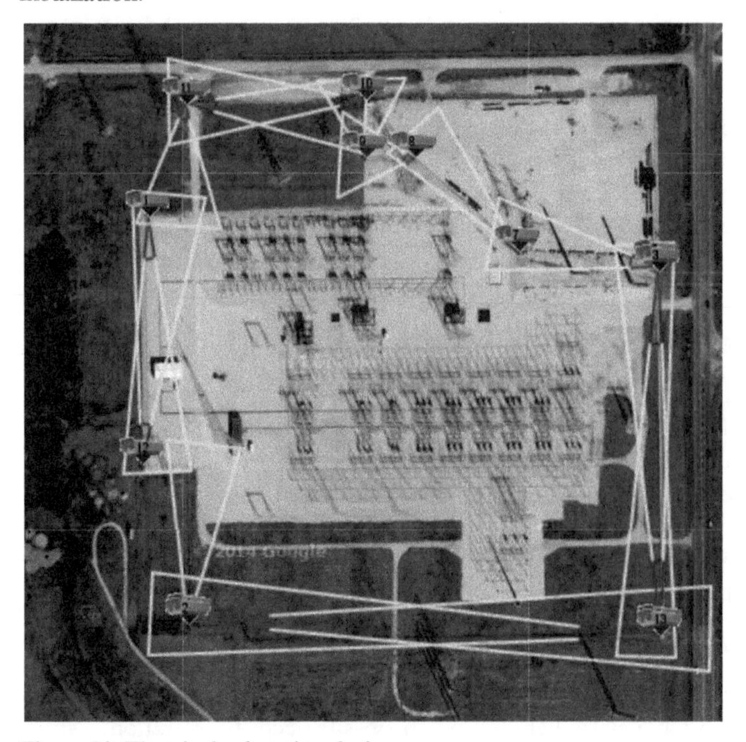

Figure 23: Electrical substation design

Challenges

- Remote location made installation costly and difficult
- Irregular perimeter created several potential blind spots

- Multiple site assets (transformers) made it impractical to use an asset or area coverage scheme

Solutions

Cost control. The use of long-range cameras means fewer units are required. This translates into fewer mounting poles, less trenching, and other savings in system infrastructure. In addition, as many of the poles as possible carry two cameras, further reducing infrastructure costs and headaches.

Reliable alerts. Longer-range cameras monitor the perimeter, providing coverage on both sides of the fence line. This allows security managers to set up a from/to detection zone on the outside of the fence, improving alert reliability.

Redundant coverage. Facing cameras cover each other's blind spots while watching the same area. If one loses power, for example, the other will still provide coverage.

Adapted to site. Shorter-range cameras with wider fields of view handle the blind spots created by the site's irregular shape.

Application: Transportation

Airport: Perimeter

Coverage: Perimeter
Monitoring: On-site Security Operations Center

Airport runways are by necessity long and exposed. They range from 2,700 ft (823m) for planes with fewer than 10 passenger seats to over 8,000 ft (2,438m) for wide-body commercial jets. In this design the use of fewer cameras with very long detection ranges significantly lowers costs by reducing the amount of mounting poles and related infrastructure needed.

Figure 24: Airport perimeter

Challenges

- Long fence lines along runways make coverage difficult and costly
- Potential intruders can enter far away from the terminal to avoid detection, especially at night
- Lots of distracting movement on or near the runway
- Potential blind spots immediately in front of cameras and at fence corners

Solution

Extended reach. Very long-range cameras (600m detection range) cover fence lines with the minimum number of units and poles for greatly reduced equipment and installation cost.

24-hour smart detection. Smart thermal video technology delivers reliable motion detection day and night, without artificial lighting. The cameras are unaffected by runway lighting, and are programmed to ignore animals, birds, airplanes, and authorized service vehicles.

Total coverage. Each camera is located 515m from the next so it can cover its distant neighbor's blind spot. Facing cameras on the short sides of the runway area pick up each other's blind spot as well as the corners of the perimeter.

Application: Transportation

Light Rail: On-Track Object Detection

Coverage: Area
Monitoring: On-site Security Operations Center

Railroad platforms and nearby tracks are full of movement: trains, blowing litter, even small animals. Any of these objects could set off an alarm in a normal surveillance system. Smart thermal video cameras in this design block them out without impairing detection of a human who has fallen or walked onto the tracks. The purpose of this solution is to detect people on the tracks so that with sufficient advance notice the train can be stopped in time to save lives.

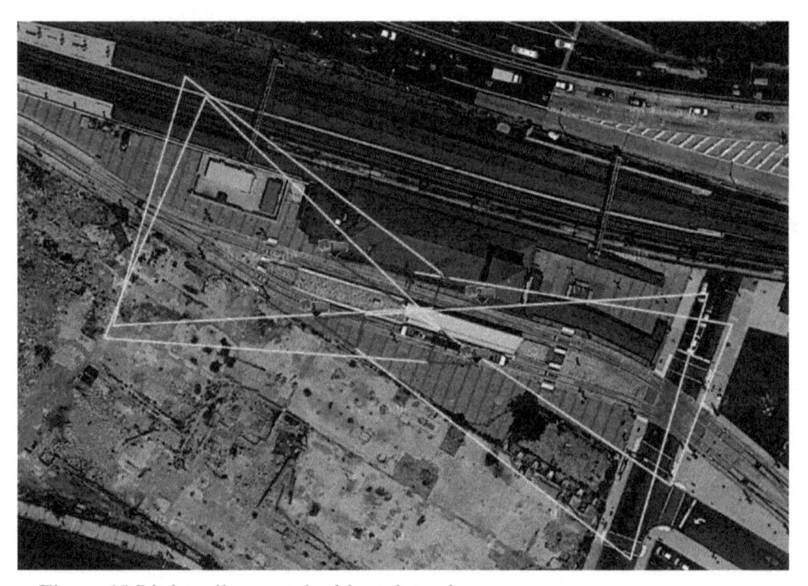

Figure 25:Light rail on-track object detection

Challenges

- Engineers and station personnel need instant notification when someone is on the tracks
- Monitoring tracks around the station requires accurate motion detection that ignores lights, signals, and other normal activities
- System must recognize when a person leaves a safe zone

Solution

Total coverage, minimal equipment. Only four cameras are needed in most cases to cover platforms and tracks. Overlapping patterns prevent blind spots and enable detection throughout the passenger-accessible areas of the station.

Reliable, accurate alerts. Criteria such as size, speed, and direction of movement are programmed into the detection rules so that only human figures trigger an alarm. Lights, blowing debris, small animals, and trains are ignored.

Safety Zones. Cameras are mounted directly above the warning stripes on the passenger platforms, allowing detection of any movement from safe zones into danger areas.

Application: Corrections

Corrections Facility

Coverage: Perimeter
Monitoring: On-site Security Operations Center

Like most prison complexes, this is a rectangular facility enclosed by long walls, with manned watchtowers at each corner. It requires 24-hour security with no gaps to prevent contraband from being thrown into the facility and keep convicts from escaping. The smart thermal video camera-based security system covers the entire facility with a minimum number of cameras, without a separate infrastructure.

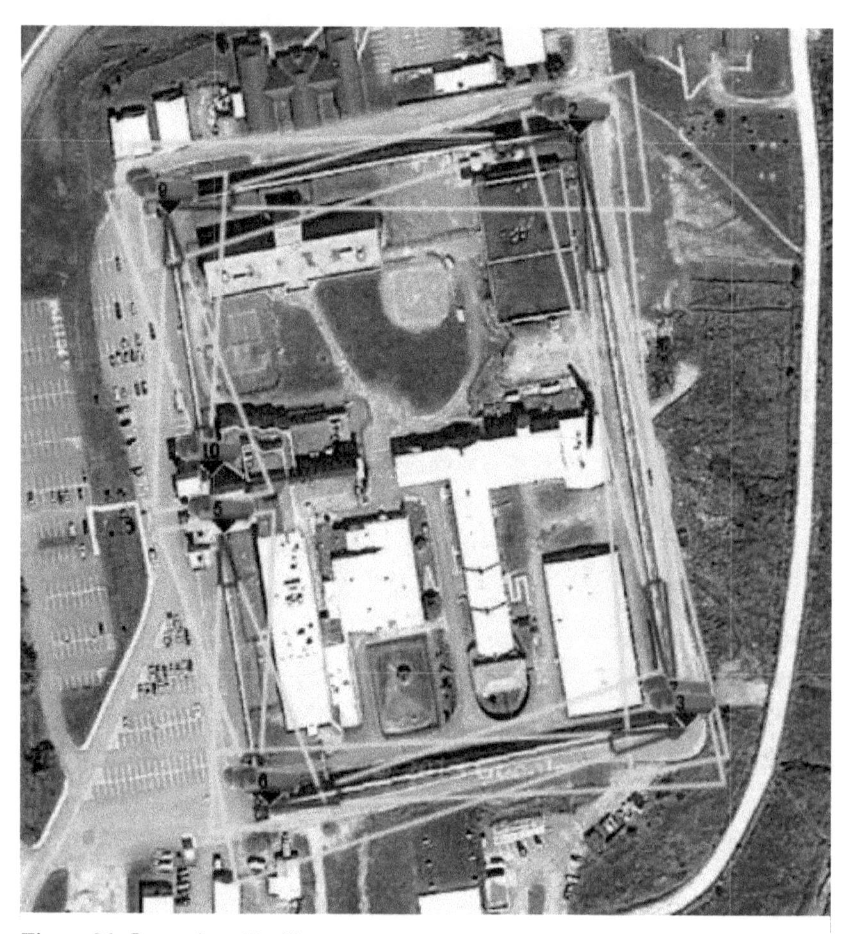

Figure 26: Corrections Facility

Challenges

- Corrections facilities need total coverage with no gaps to detect unauthorized people entering or leaving the restricted area.
- Erecting poles to mount cameras is difficult, requires costly infrastructure
- Prison sites are often large; cameras must have long detection ranges

Solution

Redundancy and blind spot coverage. Facing cameras on all four walls cover each other's blind spots, and provide redundant coverage of crucial areas without increasing infrastructure costs. Cameras on either side of the main gate watch a potentially major blind spot.

Watchtower mounting reduces costs. Long-range cameras mounted on the watchtowers at all four corners cover the full length of each wall. No new mounting poles or wiring for power or communications are needed, cutting new infrastructure costs to zero.

Long-range, 24-hour detection inside and outside walls. Smart thermal cameras deliver accurate, reliable intruder detection even in the dark. Their volumetric coverage detects activity on both sides of a wall, alerting guards to unauthorized prisoner movement inside as well as to accomplices attempting to throw contraband over the wall from outside.

Application: Commercial Sites and Buildings

Data Center

Coverage: Perimeter
Monitoring: On-site Security Operations Center

The use of long-range smart thermal video cameras allows this data center facility to limit pole locations, trenching, and wiring, greatly reducing infrastructure and installation costs. The building's irregular shape requires some overlapping of camera coverage in order to eliminate blind spots.

Figure 27: Data center

Challenges

- Data center requires 24-hour vigilance for multiple buildings
- Several blind spots along the perimeter due to site configuration
- Perimeter is not rectilinear; unrelated facilities cut into its rectangular profile at opposite corners

Solution

Accurate alerts, PTZ control. Smart thermal video cameras deliver clear images and reliable, accurate intrusion detection day or night. They can also direct PTZ cameras to the exact location of a suspected intrusion for clear, close-up images.

Overlapping coverage. Facing cameras cover each other's blind spots. They also provide redundant security for all sides of the facility.

Detection range choices cut costs. The design matches the dimension to be covered to camera detection ranges. Using a long-range camera on a long, uninterrupted border eliminates the expense of setting up multiple mounting poles. Less expensive shorter-range cameras provide 100% blind spot coverage where other buildings cut into the facility's profile.

Car Dealer's Lot

Coverage: Area
Monitoring: Central Station Monitoring Service

Smart video security systems don't have to be elaborate. The system at this car dealer provides full area coverage of two outdoor storage lots with just two cameras. Its purpose is to protect against intruders intent on theft and vandalism of the dealer's major asset, the inventory of cars.

Figure 28: Car Dealer's Lot

Challenges

- Car lots need night-time monitoring to protect against vehicle theft and vandalism
- Perimeter coverage is less important than detecting motion toward and among new and used vehicles stored on the lots
- Dealers don't have budgets for adding expensive infrastructure
- Most dealers don't employ a security staff

Solution

Security is active at night. Smart thermal video cameras deliver reliable intruder detection at night, without the need for even artificial light. Unlike conventional surveillance they provide active alerts of suspicious activity.

Area coverage blankets the lots. The cameras will discriminate between humans and other moving objects. They can also be set up with from/to zones, so movement from surrounding areas into the lots get priority tracking.

Simple yet sophisticated. Only two wide-angle (90°) cameras cover both lots. They are mounted on buildings with ready access to electrical and communications wiring, eliminating the need for additional infrastructure.

Professional monitoring. Alerts are forwarded to an off-site monitoring service for appropriate action.

Scrapyards 1 & 2

Coverage: Area
Monitoring: Central Station Monitoring Service

Scrapyards are busy during the day, but need security to protect their often valuable contents from theft and vandalism after dark. Fences are costly, especially for larger yards, and will not stop determined thieves. But by using area coverage scrapyard #1 (Figure 29) needs just one smart thermal video camera to detect intruders throughout the yard, while #2 (Figure 30) gets complete coverage with only two.

Figure 29: Scrapyard #1

Figure 30: Scrapyard #2

Challenges

- Night-time monitoring especially important
- Assets in the yards need protection from theft and vandalism; perimeters are less important
- Detecting heat is very important for the early detection of fires
- Cost is an issue; can't afford hiring security guards

Solution

Detects intruders at night. Smart thermal video cameras deliver reliable detection and accurate intruder alerts even in total darkness.

Total asset security. Area coverage from smart thermal cameras blankets the scrapyard's storage area(s), and can be set up to alert security when an object of the designated type moves into those areas from a specified direction.

Fire detection. The heat-sensing cameras can act as early-warning systems for fires, always a concern for scrapyards.

No poles or elaborate infrastructure. The cameras are mounted on buildings where they have easy access to power and communications lines, eliminating the need for costly installation of separate poles and wiring. There's no need for artificial lighting, reducing operating costs. And the detection and alert functions are so accurate they can be monitored by an off-site central service.

6

New Directions in Video Security

Smart video technology is transforming outdoor site security. Yet we have only just begun to realize its potential. In this chapter we'll look at new applications on the horizon, and assess emerging technology that promises to propel smart video into markets beyond the tightly defined worlds of corporate and government security.

We begin by projecting how the ongoing evolution of smart thermal video and related technologies will affect existing site security applications. Then we'll take on the riskier task of looking into the future to imagine how smart video, thermal and visible, will expand from this base to open up different and much larger markets for the security industry.

Developments in Technology

Smaller, faster, cheaper is the formula for progress in the electronic age. Over the past dozen years smart thermal video has been a prime example of this drive for improvement. During that time

- Camera size shrank by more than 70%
- Imager resolution increased significantly
- Image processors became 100X more powerful
- Nuisance alerts dropped to .01% of previous levels
- Camera power requirements fell to a small fraction of their original values
- Thermal imager prices plummeted, from over $8,000 to under $100 for the least expensive model

Related areas of electronics have also seen significant progress, such as battery technology and wireless communications. These developments are now having an impact on smart thermal video for outdoor site security, accelerating the technology's growing acceptance by security professionals.

Innovations in Smart Thermal Outdoor Security Video

Now that smart thermal video cameras are smaller, cheaper, have more accurate detection and clearer images, and are less power-hungry, they are even better positioned to bring solid, reliable security functions to problematic outdoor sites. We'll look first at the advantages of cutting two cords: communications and power.

Cameras without Wires

Give a camera wireless communications capability and it becomes less expensive to install. It is less expensive because it sets up quickly, with no communications wire to run.

Wireless networking also gives a site's security director more options. For example, when the signal is sent down a wire it goes exclusively to the on-site Security Operations Center (SOC). But a wireless signal propagates throughout the site. Authorized security officers on patrol can access it with tablet or laptop computers – or even cell phones – effectively becoming mobile SOCs.

It's possible for determined intruders to intercept a wireless signal, of course, but it won't do them much good if the signal is encrypted. Encryption prevents trespassers from even seeing the video in the bitstream. Only properly equipped staff can see the imagery – and receive the intrusion alerts that travel with it.

Power Play

Our wireless paradigm is not restricted to signal transmission. The next logical step is a camera with no wires at all, not even to supply electric power. As the power requirements of smart thermal video cameras continue to drop, it becomes feasible to run them on solar cells with battery backup.

In this scenario each camera on a site would get its power during the day from a solar cell that is also recharging a battery pack. At night it would switch to the battery pack for power. The result: autonomous cameras, with zero reliance on hard-wired connections.

- Totally untethered cameras have three other benefits.
- They permit the installation of smart thermal video security systems in areas where power lines are not readily available, or over terrain where it's difficult to string power and signal cables.
- With no wires to impede installation, cameras can be placed in locations where they are less conspicuous.
- A wireless approach greatly reduces installation and infrastructure costs, making it an attractive option even in locations where it's possible to run wires.

Metadata Integration

Smart thermal video does not travel through a network alone. Cameras also transmit metadata that describes such parameters as

- Time and date stamp
- Site geo-registration
- Camera location
- GPS locations of objects of interest

and other information important to a security operation. This data can be used in real time or stored with the video images for later review.

If the SOC is using a physical security information management (PSIM) system to monitor an outdoor site, camera metadata provides valuable input. A PSIM system integrates information from multiple sources – sensors, radar, video – into an overview of a site's total security operation. Law enforcement uses a similar approach, called a Fusion Center. The addition of smart thermal video metadata enables either system to monitor a site more effectively and efficiently.

Central Station Monitoring

The breakthrough application for smart thermal video security, however, is central station (CS) monitoring.

CS monitoring is the industrial version of home protection systems. It combines on-site intrusion detection with remote monitoring by professionals in a central office. CS monitoring promises to permanently change the face of outdoor site security.

Unlike a PSIM, which is structured to allow local review of alerts before responding to an apparent intrusion, CS is binary. There is no local review, and thus no dependence on the vagaries of individual judgment. Either the apparent intrusion goes to the monitoring station or nothing happens.

CS monitoring works on the basis of a script. Security system designers construct a well-defined set of guidelines on what constitutes a breach. If a suspicious activity conforms to that script, the monitoring agency will call local security or law enforcement officials to investigate.

This approach only works when false alarms from the system are held to a minimum. Smart thermal video security, with its virtual absence of nuisance alerts, is the first surveillance and security technology accurate enough for CS monitoring of outdoor areas.

A CS monitoring approach removes the burden of maintaining a staff of trained monitoring specialists on location. The personnel are provided by the remote CS service, fully trained and always on duty when needed.

We expect rapid growth in the number of outdoor sites that employ CS monitoring services. In fact, there are already successful CS monitoring companies offering services customized to smart thermal video security installations.

Market Expansion: Commercial to Consumer

So far every smart thermal video application in this book has been for an industrial, governmental, or public outdoor site. Residential security has been conspicuous by its absence. That may be about to change.

There are several good reasons why homeowners don't install surveillance or security cameras to cover their property. The biggest reason to date has been cost: smart thermal video cameras are more expensive than other sensors.

Today, thanks in part to the steep price drops mentioned at the start of this chapter, smart thermal cameras are being installed in high-end residences. Once their prices reach the sub-$100 range, however, you will see average homeowners adding them to their CS-monitored indoor alarm systems for coverage of their outdoor property.

Consumers will understand that it's far better to detect an intruder approaching your house from your backyard than to wait until he or she is already inside your home. Once that capability becomes truly affordable, the market for smart thermal video cameras will expand exponentially.

Lifestyle Management

What if security system installers and integrators were selling not just a security system, but a tool to manage activities around the house while it provided security?

Lifestyle choices are not subject to the same rationale as security systems. If a system helps the homeowner live the way he or she wants to live, and reduces the risk of accident, the value calculation becomes completely different.

Pool Safety and Beyond

For example, homeowners with a backyard pool may respond to the idea that a smart thermal video camera lets them keep an eye on poolside activity. Thermal video's immunity to reflections off the water makes it possible to find out if kids are swimming without adult supervision. When the camera sends an alert about activity in the pool, the homeowner can access the video on a tablet or cell phone, and use a loudspeaker built into the camera to warn the offenders to stop swimming until an adult arrives.

These and many other lifestyle applications change smart thermal video security cameras from a purchase that a homeowner will begrudge having to make into a great way to make life more convenient. It's not security – it's a form of lifestyle management, or (in organizational terms) operational effectiveness.

In this context the system's ability to tell you if there's a burglar on your property in the dead of night is almost a fringe benefit.

How far away are applications like this? They could be closer than people think. Right now at least one manufacturer is developing a smart thermal video camera the size of a small camcorder with built-in wireless networking and built-in loudspeaker, specifically designed to mount on the side of a house or elsewhere within a residential property.

When this catches on, we expect smart thermal video security to expand very quickly from corporate and government applications into the consumer and personal products market. The experts who sell and install such systems will certainly profit from such a market extension. More importantly, however, those whose assets they protect will benefit as well, especially if those assets are small children rather than large transformers.

Smart Video for Indoor Applications

Smart thermal video is the most advanced technology available for 24-hour detection of human intruders. Why, then, has its adoption for outdoor security use taken so long? And why has it so far failed to make any headway in indoor security systems?

In the case of outdoor security the major impediment was the engineering challenge. Creating a smart video-based system that combined reliable detection with very low nuisance alerts designed to excel in the tough outdoors proved to be a very difficult task.

Cost was also a factor in the early days, when thermal cameras were very expensive. That barrier has been mitigated or eliminated as thermal camera prices dropped and thermal systems demonstrated the ability to cover a site with fewer cameras (and thus less infrastructure) than their visible light-based counterparts.

Cost vs Potential Loss

With indoor systems, however, the biggest barrier was always cost. It still is. Indoor systems generally rely on PIR (passive infrared) motion detectors, which cost only a few dollars each. Even an inexpensive visible-light video camera costs far more than a PIR detector today.

Two new developments promise to change that equation. On the cost front, the ultimate solution would be a smart video-based solution at the same price point as a motion detector alarm system.

This is approaching the realm of possibility with less expensive thermal imagers as well as visible-light cameras. Without the challenges of an outdoor space, smart cameras of both types can do a credible job of detecting motion in a warehouse, factory, or office environment. They would be used mostly after hours, but such spaces often have some ambient lighting even at night, and supplemental lighting would be easy to provide.

A smart security camera system at the same price as a conventional PIR motion detector system would be a game-changer. Motion detectors are notoriously finicky and temperamental, and have limited range. Functionally all they can do is send an alarm that something moved across their path. Pixels in a video image, on the other hand, are like a blanket of sensors across a wide area. No other sensor can match a smart camera for the volume of information it provides, or for finding targets and disqualifying non-threats.

Stopping Warehouse Theft

As less costly smart video indoor security systems begin to appear, we can expect that firms with a lot at risk will start installing them as replacements for their current alarm systems. For example, warehouse thefts are a growing problem, and the stakes are huge. One example will show why warehouse operators need better security.

On March 13, 2010 a group of Florida men pulled off one of the biggest heists in history, stealing between $50 and $100 million worth of pharmaceuticals from an Eli Lilly Company storage warehouse in Enfield, CT. They evidently had inside information about the facility's alarm system, and got in through a hole they cut in the roof at a spot where they could avoid motion detectors.

Once they had disabled the alarm system from inside, they loaded approximately 49 pallets of pharmaceuticals onto their truck and left. The theft was only discovered the next afternoon.

This is precisely the kind of situation where smart video security would make the difference. All that's needed is the right system. If the cost is reasonable, companies with this much at risk will have no compunctions about replacing a motion sensor alarm with a much more effective smart video security system that's close to the same cost. We look for smart video to make major inroads in the indoor commercial security market.

Residential Potential

Once smart video indoor security systems reach cost parity with motion sensors, there's no reason why homeowners won't want one too, to better protect their houses while they're away. The advantages are obvious: more reliable detection, fewer false alarms, more information. The idea that the homeowner can see what's happening, instead of just getting an alert, will also have great appeal. Ultimately, this will provide a video verified alarm which police departments will give priority over other alarms.

We're talking about a new world for smart video. It is not far off.

Summing Up

We end as we began. In the opening pages of this book we pointed out that the ultimate goal of a security system was to detect potential intruders before they could do any damage. Our purpose was to show how smart thermal video security achieves that goal for outdoor sites, a feat not possible with any other approach, and to give readers guidelines on creating successful smart thermal video security systems.

In this chapter we have taken that thesis much further.

- We foresee the technology moving from large commercial, governmental, and public sites to private residential use.
- We are forecasting indoor applications for smart thermal video security systems.
- We are even anticipating the possibility of applying in-camera processing to visible-light cameras for indoor use, thus lowering the cost of security for industrial and residential users alike.

Whatever the application, it's clear that smart video security technology is poised to assume a major role in protecting sites, houses, property, and people. It is making the world a safer place.

About the Authors

John Romanowich

John. Romanowich is the president and CEO of SightLogix, a leading smart camera supplier for outdoor security applications. He has two decades of success in building technology companies, having held management positions in video technologies with Intel, IBM and the Sarnoff Corporation. He also co-founded Pyramid Vision Technologies, a pioneer in intelligent video surveillance acquired by L3 Communications. John is a board member of the Security Industry Association (SIA) and serves as the Chairman of SIA's Critical Infrastructure Protection Working Group. He is also the chairman of SIA's Perimeter Security Standards Subcommittee and a member of the RTCA Airport Perimeter Intrusion Detection Standards Group.

Danny Chin

Danny Chin holds 46 patents in video processing. He is currently the Vice President of Engineering and a founder of SightLogix. Danny was previously the Director of Advanced Development at DIVA Systems where he was the chief architect of three generations of video-processing-architectures. Prior to that, Danny was with Sarnoff Corporation for 19 years, where he led many hardware and software development teams engaged in sophisticated video and image processing development.

Thomas V. Lento

Tom Lento is the author or co-author of six recent business books and many articles on business and technology. He has led advertising, PR, and marketing communications programs in agency and corporate settings, and ran a consultancy providing strategic PR and corporate communications services to major technology companies. Dr. Lento has also served on the faculties of universities in the US and Japan.